EARTH SCIENCES IN THE 21ST CENTURY

WEATHERING

TYPES, PROCESSES AND EFFECTS

EARTH SCIENCES IN THE 21ST CENTURY

Additional books in this series can be found on Nova's website under the Series tab.

Additional E-books in this series can be found on Nova's website under the E-books tab.

EARTH SCIENCES IN THE 21ST CENTURY

WEATHERING

TYPES, PROCESSES AND EFFECTS

MATTHEW J. COLON
EDITOR

Nova Science Publishers, Inc.
New York

For permission to use material from this book please contact us:
Telephone 631-231-7269; Fax 631-231-8175
Web Site: http://www.novapublishers.com

Additional color graphics may be available in the e-book version of this book.

Library of Congress Cataloging-in-Publication Data

Weathering : types, processes, and effects / editor, Matthew J. Colon.
p. cm.
Includes bibliographical references and index.
ISBN 978-1-61324-280-3 (hardcover)
1. Weathering. I. Colon, Matthew J.
QE570.W44 2011
627'.5--dc22

2011011434

Published by Nova Science Publishers, Inc. † New York

CONTENTS

PREFACE

Weathering is an important phenomenon of the geochemical cycle as it contributes to the relief formation. It corresponds to a general term applied to physical and chemical changes suffered by rocks as a consequence of their exposition to different conditions of humidity and temperature. In this book, the authors present topical research in the study of the types, processes and effects of weathering. Topics discussed include the use of weathering indices in rock research; geomorphic processes in the last glacial age; understanding chemical weathering in affecting the Earth's surface; weathering of dimensional granite stones used as cladding and the weathering process and desert pavement development.

Chapter 1 – One of the practical applications of research into the types and processes of rock weathering is in the field of rock art studies, where it plays a key role in two areas. First, it is among the most crucial indices in efforts of estimating the age of rock art, most especially that of petroglyphs. In this area, weathering is arguably the most promising variable in the 'direct dating' methodology that has been developed in recent years. The reasons for the failures of alternative dating methods are explored, leading to the proposition that weathering and related features offer the most reliable basis for future work in this field. Some of this new methodology is discussed within the overall context of the considerable difficulties generally experienced in rock art dating. The second role of weathering in rock art research is in the field of conservation and preservation. Here the various processes of weathering are of considerable consequences in developing appropriate strategies for the demands of cultural resource management practices.

Chapter 2 – A lot of studies on the restoration of the past geomorphic processes on the slopes of the Japanese mountains have been executed mainly for the glacial and periglacial features by the Japanese researchers: Koaze et al. (1974), Ono and Hirakawa (1975), Yanagimachi (1987), Oguchi (1992), Koizumi (1992) and others. Because the slopes in these remaining features are limited to those located at the altitudes that are higher than the glacial forest line, there have been many reports on the subject of the Japan Alps, Kitakami Massif, and high mountains in Hokkaido. On the other hand, a great number of slopes that have not been examined for the past geomorphic processes exist in high zones such as the middle relief mountains of central Japan where a range of the glacial and periglacial features does not exist. It can be said that the clarification of the formation history and process of such unexamined slopes provides the essential means for investigating totally both the history of geomorphic processes and the history of environmental change in the Japanese Islands.

Chapter 3 – In managing our hydrological systems, increasing use is being made of information generated from studies held for understanding the controls on chemical weathering due to its importance in affecting the evolution of the Earth's surface, shaping landscapes, determining nutrient supply to ecosystems, and regulating global chemical cycles. A lot of recent studies have been realized to determine the parameters controlling the denudation, chemical weathering and physical erosion under different climatic conditions. In order to estimate chemical weathering fluxes affecting the drainage basins, models utilizing the Na, Ca, K, Mg and total dissolved load concentrations have been applied to obtain fractions coming from dissolved rocks, after corrections for river waters of the atmospheric inputs, dissolved solids from ion exchange sites in clays minerals, dissolved solids due to changes in biomass and anthropogenic inputs, among other factors. Since uranium is among the main elements contributing to natural terrestrial radioactivity, several descendants of the mass number (4n+2) ^{238}U decay series have also been utilized in studies focusing weathered soil covers and hydrological resources in drainage basins. The ^{238}U decay series finishes at the stable ^{206}Pb and the radionuclides ^{238}U and ^{234}U have been used to evaluate the chemical weathering rate in soil profiles and hydrographic basins based on the fact that ^{234}U is preferentially mobilized to ^{238}U when rock weathers in the alteration front. Measurements of the ^{234}U/^{238}U activity ratio (AR) in rocks, soils and waters have allowed the calculation of the solution coefficient for the uranium characteristic of the region, that allows evaluate the time necessary to weather 1 m of rock under actual climatic conditions. This chapter summarizes some studies realized in a soil profile occurring at the giant Paraná sedimentary basin, Brazil, which have taken into account the presence of natural radionuclides belonging to the ^{238}U decay series.

Chapter 4 – Granites have been used as building stones for centuries since they are very durable materials and allow different finishes such as polishing, flaming, hammering and so on. When they are placed in a building, their surface is exposed to a wide range of weathering agents, which can produce damage, and consequently changes, in their aspect. The resultant decay depends on their petrographic and physical characteristics, and also on the environment (Del Monte et al., 1996; Chabas y Jeanette, 2001; Esbert et al., 2004). Weathering forms in historical granite buildings are well known, such as detachment, discolouration, grain disaggregation, crusts formation and/or biological colonisation. However, weathering forms of new panels in existing buildings are less identified.

Chapter 5 – There are arguments in the literature about the desert pavement mechanism and processes. One argument suggests that desert pavements result from wind deflation, the other argues that they result from mechanical weathering and concurrent soil formation. In the northern United Arab Emirates (UAE) the two types of desert pavement are found. The desert pavements which develop by wind deflation are found in depressions between sand dunes, and desert pavements which develop by mechanical weathering and soil formation are found on fluvial surfaces (terraces and alluvial fans). The first type occurs as a result of the winnowing removal of fine material from the surface, exposing underlying gravels. In the UAE these are characterised by a fine texture of small stones (cm size) protecting soil beneath. The second type involves mechanical weathering of clasts in situ at the same time as soil formation takes place. The stones are modified by weathering processes, which results in the fragmentation of the clasts. Windblown fines are trapped within the fabric and washed through the pavement surface. Wetting and drying and perhaps bioturbation reorganise the clasts to produce smooth interlocked pavement surfaces.

The rate of desert pavement development is different between the two types and may differ within the same type according to the weathering behaviour. For instance, for the deflation type, the deflation rate may be similar on surfaces of different age, producing pavement properties that may not be age-related. They can develop within hours (24-48), depending on the wind speed. For the mechanical weathering type, different lithologies or a different climatic regime may influence the rate of development, but generally the pavement properties (type of rock and weathering mechanism) will show clear age-related properties. On deflation pavements fragment angularity may be less well defined than on mechanical weathering pavements, and may shows a decrease in angularity on older pavements. Clasts on mechanical weathering pavements become progressively more angular with age and smaller in size in relation to the subsurface material. However, salt weathering and rock varnish may smooth the edges on older surfaces. Furthermore, microtopography on the pavement surface, related to original deposition becomes progressively obliterated. Age-related differentiation on both types of pavement may be undertaken through a consideration of soil maturity (colour and texture and profile development).

In: Weathering: Types, Processes and Effects ISBN 978-1-61324-280-3
Editor: Matthew J. Colon, pp. 1-67

Chapter 1

THE USE OF WEATHERING INDICES IN ROCK ART RESEARCH

Robert G. Bednarik*

International Federation of Rock Art Organizations (IFRAO), P.O. Box 216, Caulfield South, VIC 3162, Australia

Abstract

One of the practical applications of research into the types and processes of rock weathering is in the field of rock art studies, where it plays a key role in two areas. First, it is among the most crucial indices in efforts of estimating the age of rock art, most especially that of petroglyphs. In this area, weathering is arguably the most promising variable in the 'direct dating' methodology that has been developed in recent years. The reasons for the failures of alternative dating methods are explored, leading to the proposition that weathering and related features offer the most reliable basis for future work in this field. Some of this new methodology is discussed within the overall context of the considerable difficulties generally experienced in rock art dating. The second role of weathering in rock art research is in the field of conservation and preservation. Here the various processes of weathering are of considerable consequences in developing appropriate strategies for the demands of cultural resource management practices.

Introduction

Rocks form through many different processes, but once exposed at the Earth's surface they deteriorate through the universal process of weathering. While the rate of rock formation differs enormously, from rapid crystallization of volcanic rocks to metamorphic modification that can extend over geological epochs of many million years, the courses weathering takes are of a more limited spectrum of time. Nevertheless, there are considerable differences in the duration a given amount of weathering will take, depending on the rock type and the ambient conditions of moisture, climate, pH and so forth. The lithosphere's principal agent of rock weathering is water, most especially derived from precipitation. Water occurs in pores or cracks within rocks in gravitational, capillary, or hygroscopic form (as thin films on grains), or it is held in chemical combination in some rock-forming minerals. Apart from oxygen, the two most abundant chemical elements in the lithosphere are silicon and aluminum, and the most abundant minerals are quartz and aluminosilicates of various kinds. As quartz is almost immune to weathering in most surficial conditions, chemical weathering is principally concerned with aluminosilicate chemistry, and the principal products of rock weathering are the minerals they decay to, clays and various oxides. These processes are considered in other contributions to this volume, but in general, they are essentially solution, hydration, hydrolysis, ion exchange (including chelation), oxidation, reduction, and carbonation.

Chemical weathering is almost invariably conditional upon the presence of water; even oxidation of minerals by gaseous oxygen appears to require water as an intermediary agent (Keller 1957). Such weathering is usually by means of very complex processes that are frequently rendered more intricate by the ability of many of their own products to accelerate them. For instance, a byproduct of the oxidation of pyrite is sulfuric acid, which readily reacts with numerous minerals; or hydrogen clays, themselves the result of weathering, will induce hydrolysis with their hydrogen ions. Atmospheric water travels within the surficial weathered and porous zones of the lithosphere by means of gravity, capillarity or heat-induced circulation. Although atmospheric precipitation can range from pH 3 to pH 9, most weathering reactions occur in the acidic range.

This paper focuses on the quantification of the products of these processes as they are of relevance to the scientific study of rock art and archaeology, and of anthropic rock surfaces generally. The morphological effects that are perhaps of the greatest significance to the rock art researcher are surface retreat, saprolithization (the initial decay phase of rock), and the development of weathering rinds and patinae.

It is well known that rock surfaces retreat with time, be it by solution (Acker and Bricker 1992; Busenberg and Clemency 1976; Lin and Clemency 1981; Oxburgh et al. 1994; Rimstidt and Barnes 1980; Williamson and Rimstidt 1994), physical wear, or a combination of both. But to render weathering processes useful for rock art age estimation ('dating'), ways need to be found to quantify them: at what rates, relative to rock type and ambient environment, do they occur, and how can they be measured and calibrated? Much the same applies to the potential ability of countering rock weathering in the quest to preserve rock art or other immovable cultural heritage made of rock: the processes of deterioration need to be understood thoroughly before intervention to arrest them can be considered.

Among common rock-forming minerals, susceptibility to chemical weathering decreases in the order: calcite, dolomite, anorthite, olivine, augite, pyrite, magnetite, hornblende, biotite, albite, plagioclase, orthoclase, microcline, epidote, muscovite, quartz. Solution rates of minerals in typical water compositions vary enormously under laboratory conditions, from 1 mm/31 days for calcite (Chou et al. 1989) to 1 mm/34 million years for crystalline quartz (Rimstidt and Barnes 1980). In a more practical sense, the sequence is better illustrated through the rate of surface retreat as determined from monumental masonry of known ages. Such retreat varies greatly according to local rock composition and climate, but the rock art student needs to be familiar with empirically determined ranges for common rock types, at least as a rough guide. These are, in millimeters per one thousand years, under central European climatic conditions:

Sandstone	5 – 50
Limestone	2 – 20
Schist	1 – 10
Marble	0.4 – 5
Dolomite	0.3 – 2.5
Serpentine	0.25 – 2.5
Diabase, porphyry	0.2 – 2
Gabbro	0.1 – 1.5
Diorite	0.1 – 1
Quartzite	0.1 – 0.5
Granite	0.05 – 0.2

It is therefore unrealistic, for example, to expect to find Pleistocene petroglyphs on a schist panel that was always fully exposed to precipitation, because at these rates of surface retreat (for schist 1–10 cm since the end of the Pleistocene) it is highly unlikely that any rock art could have survived under these conditions. Preservation would be significantly better in an arid or Arctic climate, where the last four listed rock types might show minimal surface deterioration over ten millennia. Indeed, petroglyphs on granite surfaces in northern Karelia are relatively well preserved after 4000 years (Bednarik 1992), while historical, dated inscriptions on schist in the Côa valley of Portugal are barely decipherable after a few centuries (Bednarik 1995). In the arid Pilbara of north-western Australia, petroglyphs have survived for many millennia on gabbro, and for several tens of thousands of years on granite (Bednarik 2002a). These kinds of observations assist the rock art student greatly in gaining an initial understanding of the potential antiquity of petroglyphs.

The second type of weathering effect after surface retreat, the formation of saprolite, is of less interest to rock art science. Decay of rock begins with the formation of a carious zone around individual grains or crystals where reaction with water occurs, and progressively leads to breakdown into detrital material. Feldspars weather to clay minerals, typically kaolinite, and in some situations even to the aluminium hydroxide mineral gibbsite. Quartz is largely unaltered in the weathering environment, while in ferromagnesian minerals such as pyroxene and amphipole weather to iron oxides and montmorillonite clays. Of particular interest is the saprolithization of sandstone as interstitial amorphous silica is leached out (it is about twenty times as soluble as quartz) and the rock disintegrates into quartz sand. This is important in appreciating the difficulties of applying thermoluminescence dating to the resulting sediments (cf. Fullagar et al. 1996; Roberts et al. 1998).

Of particular relevance to rock art science is the formation of weathering rinds, which are typically distinct zones of alteration whose thickness is a function of exposure time. If the process could be calibrated against time it would probably yield rough estimates of geomorphic exposure ages. Černohouz and Solč (1966) attempted this with basalts, claiming to obtain reliability to within 10%–20%, but there has been no adequate attempt to pursue this possibility further (Colman and Pierce 1981). Although part of their hypothesis was subsequently refuted (Bednarik 1992), they correctly recognized that weathering rind thickness is a function of surface geometry, and that this aspect is the cause of wane formation.

Weathering rinds are zones of oxidation, hydration or solution that form parallel to rock surfaces and their thickness is a function of time (Carroll 1974; Colman 1981; Colman and Pierce 1981; Crook 1986; Gellatly 1984). The growth rate of weathering rinds can be quantified for a given rock type under given climatic conditions if it can be calibrated by another dating method, but it only yields imprecise results. Surface rinds suffer from mass loss due to abrasion, erosion, frost action, *Salzsprengung* or exfoliation, which introduces a major error source for any age determination. In addition, surface shape and aspect, as well as petrological variations within the rock, also affect the process. It may be preferable to measure subsurface rinds on submerged rock, as Colman and Pierce (1981) did, examining a large sample of clasts from B-horizons of deposits. They propose a logarithmic function in the form of

$$d = \log(a + bt) \quad (1)$$

where a and b are constants, d is the rind thickness and time t can be determined. Although this is of limited use in estimating the age of petroglyphs, it does imply that quantification for dating purposes should be possible.

One of the principal difficulties in studying weathering rinds is that the need for destructive sampling normally excludes intrusive methods from consideration in the case of rock art. Destructive sampling is usually out of the question (the exception being previously fractured rocks; cf. Bednarik 2007), but the use of non-intrusive methods could be considered. The Schmidt hammer may be suitable for measuring the degree of rock surface weathering (Birkeland et al. 1979; Burke and Birkeland 1979; McCarroll 1991). This instrument was originally designed to measure the surface hardness of concrete, but has also been widely used on natural rock (Day and Goudie 1977). The Schmidt hammer has had limited use in rock art research (Campbell 1991; Sjöberg 1994; for a preliminary but unsubstantiated and inconclusive attempt, see Pope 2000), nor have there been any serious attempts so far to employ weathering rinds in estimating rock art ages in any more than the most cursory fashion. Clearly there is more research required in this area.

A relevant potential analytical method that can be useful in combination with both weathering rind and patination analyses derives from the spalling of boulders bearing rock art, be it by *Kernsprung*, insolation, kinetic impact, lightning strike or through grass, brush or forest fires. (The term 'boulder' throughout this paper refers to the granulometric definition, and is not intended to relate to its history.) Vast numbers of petroglyphs throughout the world occur on various types of rock that have been subjected to gradual size reduction by spalling. The various types of fracture can be recognized readily. Lightning generally strikes at the highest point of a rock outcrop or hill, and the impact area is typically discolored or glazed where superficial vitrification occurred. Lightning-induced fractures resemble impact fractures, commonly with a bulb of percussion or featuring radial stress lines. Spalling caused by brush fire is extremely common in some regions and results in thin flakes of up to 30 cm diameter, with a thickness of up to two centimeters, tapering towards the sharp edges due to the typically convex shape. Both insolation and fire spalling gradually remove protruding aspects and acute edges, essentially reducing rocks of any shape to a sub-spherical form (Figure 1).

Figure 1. Typical spalling scars caused by brushfire, destroying petroglyphs, eastern Pilbara, Australia.

The progressive reduction of boulders through fracture, by whatever spalling process, results in many scars, often truncated by other scars. The chronological sequence of the spalling events can be reconstructed, as schematically depicted in Figure 2. Since the analysis of the spalling sequence is within the means of a properly informed rock art recorder, it is reasonable to expect recordings of rock art on such boulders to include details of the analysis of such sequences. In other words, each relevant spalling scar needs to be identified and its chronological position relative to all other types of surface phenomena, including intentional anthropic marks (i.e. rock art), needs to be recorded. Direct dating of petroglyphs is best based on such sequencing of the traces of surface modifications, and on then locating the art within these sequences of events, the products of which can sometimes provide numerical age information directly relatable to the art.

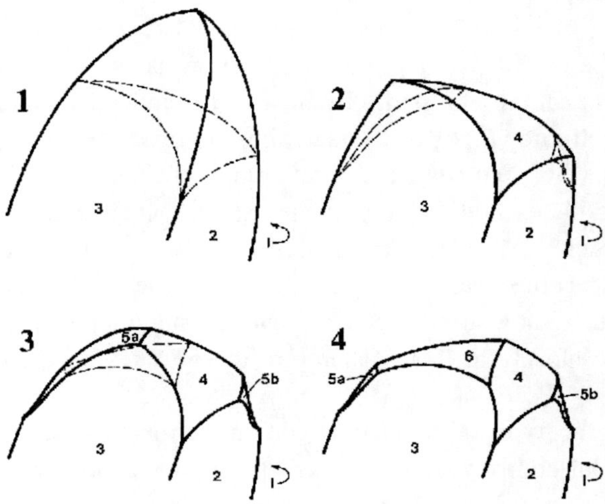

After Bednarik 1979.

Figure 2. Schematic depiction of the sequence of spalling events in the reduction of regolith boulders.

These preliminary comments imply that geometry, intrinsically involved in geological processes—from the formation of individual crystals to the laws determining weathering patterns—is particularly relevant to any quantification of processes relating to the estimation of rock art antiquity, particularly of petroglyphs. The processes attributable to insolation and fire spalling are governed by the laws of heat transfer in solids, and the same laws apply to the formation of wanes, be it at the macroscopic or microscopic level (Bednarik 1992). The tessellation of sandstone or basalt is certainly a geometric phenomenon, based on Voronoi cells. But even the formation of sandstone shelters and speleological features are governed by geometrical laws, as are the conditions of fluvial erasure of inscriptions or petroglyphs in valleys subjected to inundation (Bednarik 2009a).

Clarifying Patination

One of the key factors in chronologically sequencing the various surface facets of a boulder is that faces of different ages commonly bear different degrees of patination. The word patina, in rock art science, defines a visually obvious surface feature that differs from the unaltered rock in color or chemical composition. It is a collective, almost colloquial term for a variety of phenomena, all acquired gradually over time. Hence they are an indication of antiquity, as researchers have appreciated for centuries (Belzoni 1820). However, the use of patinae in age estimation has so far remained difficult and controversial, at least partly because of misunderstandings. For instance, there is a well-known polemic concerning the repatination of petroglyphs: if a groove has been cut into the weathering rind beneath a ferromanganese accretion, will its repatination occur at an accelerated rate relative to unaltered rock? Also, the applicability of the term to quite disparate phenomena has been responsible for erroneous views. Although most forms of patination are not, strictly speaking, weathering phenomena, some are, and others relate closely to them. Therefore these issues need to be considered here.

Unless the type of surface alteration defined as patina is identified, archaeological controversies are predictable. If the principal component of the patina was rock varnish, repatination would proceed independent of the substrate, but if the process relied largely on the oxidation of resident bedrock iron cations, then it would be affected by the state of the exposed substrate. To make this judgment it is essential to analyze the patination products responsible for the macroscopic appearance of the surface. On the other hand, archaeologists also call the weathering zone of chert, which is the result of a totally different process, a patina.

Originally, the name *patina* referred to copper carbonate or copper sulfate, the corrosion skin that develops on copper and its alloys (*verdigris*)—although the term has earlier roots and derives from the Latin word for a 'shallow dish'. It was extended to the sheen or wear polish on antique surfaces, and to cutaneous alteration of rocks or stone tools that seemed to indicate great age. Such surface skins could be the result of bleaching or leaching (e.g. of sedimentary silica; Bednarik 1980), limonite staining (Goodwin 1960), mineral accretion (e.g. by rock varnish; Engel and Sharp 1958), chemical alteration of substrate components (most rock weathering processes are candidates), and abrasion or polish (e.g. by sediment grains or biological agents). While this encompasses a great variety of causes, from the rubbing of animal bodies against a rock to the modification of optical properties through the etching of crystals, all of these processes are slow and gradual, resulting in cumulative products that represent lengthy time spans. Patinae have long been recognized as potential means of estimating the age of petroglyphs, but until recently the difficulties of quantifying and calibrating the processes involved have fostered the view that these are too intractable (but see Bednarik 2009b). Moreover, most rock surfaces experience more than one patination process, so what is simplistically called 'patination' is in fact the macroscopically visible outcome of several factors and their interplay.

Patinae can conveniently be divided into those involving the deposition of extraneous matter and endogenous alteration products (although some forms, such as the oxalate patina on marble statues, might be attributable to a combination of local and introduced substances; Del Monte and Sabbioni 1987; Lazzarini and Salvadori 1989). The form of patination most frequently encountered in the study of petroglyphs consists primarily of iron compounds, presenting themselves as the ubiquitous dark-brown coating of rock surfaces, particularly in arid and semi-arid regions. In all cases analyzed this is not the product of just one process, and it frequently combines exogenous and endogenous components. Archaeologists frequently assume this to be either an oxidation product or a rock varnish. The latter term is often misused: it should be limited to a very thin (<0.5 mm), often shiny ferromanganese skin that covers even rocks entirely free of its contributing cations, and in whose laminar, stratified deposition micro-organisms are implicated (Scheffer et al. 1963). Other deposits of similar composition and appearance occur commonly and should be described as *dunkle Rinden* (Walther 1891).

To illuminate the issues concerning the relationships of weathering depths, petroglyph depths and repatination, a graphic convention as shown in Figure 3, first published in Bednarik (1979), has been devised. This figure depicts these relationships for six core samples collected from petroglyphs at Depuch Island, Western Australia (Trendall 1964). Two petroglyphs each had been classified as being of 'fresh', 'intermediate' and 'deep' patination. In the image the mean weathering depth is compared with maximum and minimum weathering depth of each petroglyph, with the groove depth (*E*), the weathering zone thickness left beneath the petroglyph (*X*), and the position of the groove depth relative to weathering depth (*Y*). These diagrams illustrate other aspects of the relationships, for example the shape of the line representing the weathering front implies the relationship of median and mean weathering depth, and the diagrams demonstrate the lack of correlation between repatination intensity. The graphs show that, although the three samples of heavy engraving patination coincide with those of deeply weathered rock they also correspond with the deeply engraved. Furthermore they demonstrate the concurrence of the heavy (deep) engraving patination and greater distance between groove bottom and weathering front. The last-mentioned dimension steadily increases as a function of time, while it is probable that engraved groove depth decreases (although necessarily at a far slower rate), unless we postulate a complete absence of abrasive (aeolian) or other erosion. This argument introduces the possibility that of the oldest designs, only the deeply engraved remain discernible, while those of shallow depth have become unintelligible.

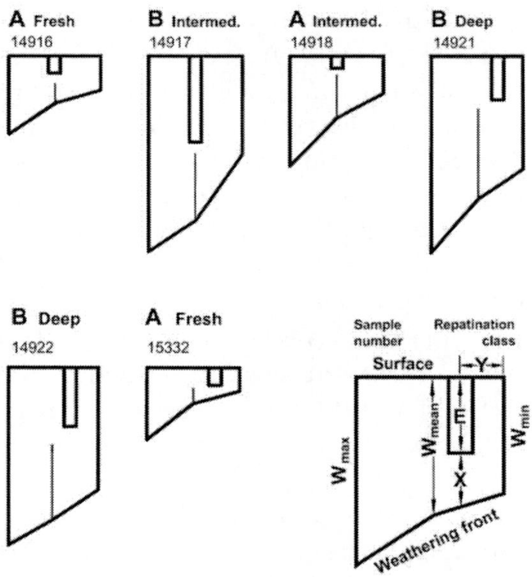

After Bednarik 1979.

Figure 3. Schematic comparison of six core samples from variously weathered petroglyphs on Depuch Island, Western Australia.

Microscopic examination of brown-patinated non-varnish surfaces usually presents evidence of a variety of processes. A surprisingly large component is clearly aeolian, especially on granites. Crusts of the *dunkle Rinden* type (a collective term for all dark rock coatings or rinds) consist of several iron compounds that are subjected to progressive and continuous modification. For instance magnetite is metastable with respect to maghemite, which in turn alters to hematite in ambient conditions at the Earth's surface, a process accelerated by solar radiation. Dehydration changes lepidocrocite to hematite, and heat can dehydrate even the much more stable goethite in the same direction, with attendant color changes. Being the most stable form of Fe_2O_3 at surficial conditions, hematite content tends to proportionally increase with time, even though it can be reduced to magnetite by high temperature. Its degree of agglomeration or particle size determines its color—the larger the particles the redder the otherwise brown deposit—but saline solutions also influence the production of red hematite. While the color is thus determined by the combination and state primarily of iron salts, especially through the taphonomy favoring hematite, other cations are also present, particularly manganese, and these crusts also comprise significant clays and aeolian detritus. Quartz, tourmaline, and other crystal grains, plant matter and even charcoal fragments have been observed in these accretions (Bednarik 2007), caked together by iron compounds and amorphous silica. They are subjected to continuous modification by rainwater, which favors the formation of distinctively 'laced' or 'terraced' morphologies. However, such extraneous deposits are usually quite discontinuous, so that exposed surfaces can be found on all but the very oldest petroglyphs. Nevertheless, in the macroscopic sense these deposits form fairly consistently as a function of time. For instance in the Australian Pilbara, an incipient film becomes evident after a fresh surface has been exposed for 30–40 years, and after about 100 years, the deposit reaches a thickness locally of 30–50 microns, or 100–150 microns after 230 years (Bednarik 2002a). Where a surface is not concealed by accretion, component minerals are subjected to conversion processes often similar to those affecting the accretions. They are here, however, much less effective as they are restricted to the weathering zone, particularly the immediate surface. The phenomenon then described as 'patina' needs to be seen as usually comprising both types of products, extraneous and intrinsic, and their complex interplay and ongoing modification regime seem to imply that these phenomena are difficult to assess effectively.

Recent work, however, suggests that there may be a simple method to 'quantify' patination. The empirical observation that, in general, repatination does proceed fairly linearly overall may be well founded. The means suggested until recently, of measuring the visual changes (e.g. by determination of reflectivity), are not expected to provide valid data, but the rapid development of computer programs to calibrate and quantify 'bulk color' with incredible precision may render patination analysis viable. The results of recent work with colorimetry have been most encouraging (Bednarik 2009b).

Other accretionary mineral deposits on rock surfaces, which are of importance for several reasons, also need to be recorded to the extent of providing the specialist researcher with basic information about the type of deposit present, and its physical relationship with the rock art. There is no arbitrary division between the thin deposits termed 'patination', and the thicker layers to be discussed here. Thin ferromanganese patinae grade into thicker deposits of similar composition, but mineral skins and crusts consisting principally of silicas, oxalates and carbonates are also common. Such mineral deposits are usually layered, can contain a great variety of datable inclusions, and most importantly, rock art may be sandwiched between two such layers. Both rock paintings and petroglyphs have been found between laminae of accretionary mineral deposits, or at least covered by them or executed on them.

Silica skins were formed through the precipitation of amorphous silica from aqueous solution of silicic acid ($Si[OH]_4$). They are rather thin (mostly <0.4 mm), of usually light color or translucent, sometimes brown or gray, and they are of vermiform surface texture. They contain a great variety of inclusions, including bacteria, pollen, fungi and algae, which may yield AMS (accelerator mass spectrometry of carbon isotopes) dates, and also numerous mineral substances of clay and silt fractions. In a classical stratigraphic study of siliceous accretions, Watchman (1996) documented the occurrence of two layers at Penascosa Panel 7 in the lower Côa valley of northern Portugal. The older is of whitish color, the younger is of a silty brown appearance, consisting largely of weathered schist detritus and platy mica fragments. An animal figure abraded into Panel 7 cuts through the white layer, whereas the superimposed brown deposit also covers the engraved groove. Consequently the petroglyph postdates the older and predates the younger crust, and its age should lie between carbon dates Watchman derived from the two.

Oxalates are not only datable through the organic inclusions they may contain, but the oxalate itself is datable by virtue of being a salt of oxalic acid, a crystalline organic acid containing ultimately atmospheric carbon and therefore ^{14}C. While the origin of the oxalates Watchman (1990) has frequently identified in northern Australian sandstone shelters remains unresolved, their dating potential has already provided useful results (Watchman 1992a, 2000). Oxalates occur as whewellite and weddellite and are generally darker than silica skins (dusty-brown to black) as well as thicker (>1 mm). They may occur in laminated deposits that include not only a variety of other minerals (e.g. gypsum) but also thin bands of pigment residues, such as ochres, which may have been applied as rock paintings. Quartz grains, clays, carbonates, and charcoal have been observed in oxalate crusts at rock art sites, and their occasionally complex laminae have yielded entire sequences of painting episodes, separated by the deposition of layers of oxalates and other minerals. The most spectacular examples of such sequences, from Arnhem Land and north Queensland, Australia, include a recorded 2.11-mm-thick micro-section from Walkunder Arch Cave that yielded ten AMS carbon isotope determinations ranging sequentially from c. 29,700 to 3300 years BP (Campbell 2000; Watchman 2000).

The analytical principles Watchman applied in the Côa valley and elsewhere were much earlier employed to study carbonate accretions (Bednarik 1981, 1984). Calcite, aragonite and dolomite are sedimentary minerals that are readily converted to bicarbonates in an aqueous solution of carbon dioxide and then mobilized and reprecipitated in a great variety of forms. Cutaneous travertine deposits up to several centimeters thick occur commonly in limestone caves, covering considerable areas of ceilings and walls. These formations may be faintly layered, and rarely they contain petroglyphs (e.g. in several caves near Mt. Gambier, South Australia), the outlines of which remain visible on the surface. The rock art as such is then sandwiched between two of the strata, having at some stage been engraved into the developing accretionary deposit. These carbonate speleothems are datable, with certain qualifications (Bednarik 1999), by several methods, especially carbon isotope and uranium series analysis. Reprecipitated calcite crusts are not limited to caves, they occur commonly in semi-arid regions in the form of a light-colored skin at the uppermost level of sediment covering rock. Deposited over other weathering and patination phenomena, sometimes over petroglyphs (Bednarik 1987), they might be datable.

The Dating of Rock Art

None of the geomorphological or geochemical methods of estimating the age of rock art offers much precision and some may seem like rather desperate measures. This raises the question: are there better, more precise methods?

The answer is that all of the many approaches that have been tried have problems. The most popular method has been determination of carbon isotope ratios of paint residues, either derived from organic components (binders, brush fibers, incidental inclusions) or from charcoal pigment. However, all rock surfaces, including accretionary materials applied as paint, contain organic material of some types. Microorganisms occur in rock kilometers below the Earth's surface, but the greatest concentrations occur near the surface, and the carbon system has long been demonstrated as being open (Bednarik 1979; Nelson 1993; Ridges et al. 2000). Therefore any sample from a rock surface is likely to yield a ^{14}C age, but this may be a random number that does not reflect the age of any rock art that happens to be present. If the paint traces were concealed by a natural mineral skin, of silica or oxalate, soon after the painting was executed, the chances of securing reasonably precise radiocarbon results increase significantly, as Watchman has demonstrated. But isolated carbon-14 dates from unprotected paint residues provide no credible age estimates, unless the paint was made with charcoal. Such black paintings, stencils or drawings are still just as contaminated by other organic matter present on any rock panel, but because the charcoal is such a dominant component in such samples the error margin is not expected to be as great. By far the most reliable carbon isotope dates from rock art are those derived from beeswax pictograms, because there is always adequate datable substance available, and the physical properties of the beeswax render it unlikely that it would have been used in any but nearly fresh condition. Therefore the time of the production of the wax, which approximates its radiocarbon age, is unlikely to differ significantly from the time the art was created (Nelson et al. 1995; Taçon and Garde 2000; Bednarik 2001a). Nelson (2000) reports radiocarbon age measurements from 137 figures at sixteen sites in the Northern Territory (Figure 4). However, beeswax rock art is limited to very few regions, having been reported only from western Arnhem Land (Brandl 1968; Chaloupka 1993), Kimberley (Welch 1995), Reynolds River, Keep River and Groote Eylandt, i.e. only from northern Australia.

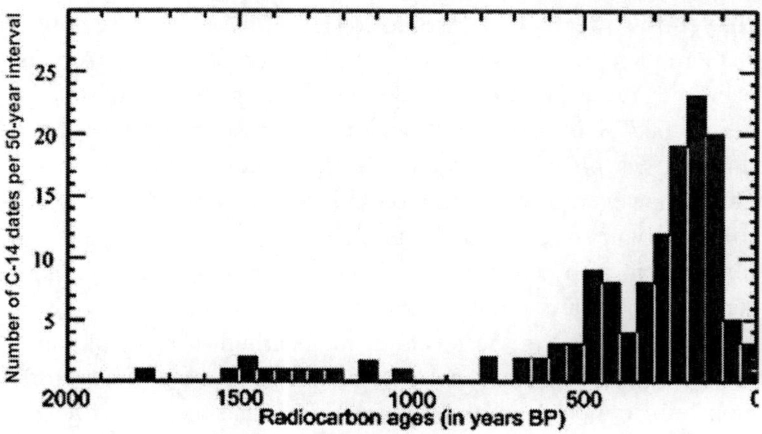

After Bednarik 2001a, from data by Nelson 2000.

Figure 4. The results of 137 carbon isotope determinations of beeswax rock art in the Northern Territory, showing the characteristics of a taphonomic logic parabola.

However, concerning the credibility of radiocarbon essays there are still other issues to be considered. Firstly, the carbon isotopes of charcoal can at best reflect the conditions at the time of assimilation, i.e. when the tree in question was alive and absorbed CO_2 from the atmosphere (Bednarik 1994a, 1996, 2000a). This must have preceded the paint application, perhaps by as much as several centuries. Secondly, the charcoal may not have been fresh at the time it was used. Charcoal often occurs in profusion on the floor of shelters and caves, some of it thousands of years old, and cases have been reported where one motif yielded two dramatically different carbon results, separated by up to tens of millennia (e.g. McDonald et al. 1990; David et al. 1999). Thirdly, it has been suggested that 'fossil wood' may have been burnt (Schiffer 1986; Fetterman 1996). Although this may seem unlikely in most cases, the wood used may certainly have been several centuries old, especially in arid regions.

These major impediments are in addition to the numerous qualifications inherent in the carbon isotope method itself. The atmospheric concentrations of carbon isotopes in the distant past are unknown to us, because they are the result of cosmogenic radiation, which fluctuates over time. The ^{14}C decay rate to nitrogen we use in all calculations, Libby's original half-life estimate of 5568 years, is incorrect; we think the true figure is 5733 years. But in any case, carbon isotope analysis yields radiocarbon years, not absolute or calendar years. To 'calibrate' them is not realistically feasible (Muzzolini

1999), yet it is often done in archaeology. But the results are not ages at all, they merely represent sets of statistical probabilities. At one sigma probability (Figure 5), the 'true' age should lie between the stated tolerance values in 68.7% of cases, but will in the rest be outside the tolerance (Bednarik 1994a). All of this applies to 'ideal' conditions, which never apply in reality, however. The assumption that the ratio of carbon isotopes in the carbon reservoir of the Earth is the result of a complete and rapid equalization is also unwarranted. Even the assumption that the ratios of carbon isotopes are only altered by radioactive decay is only an approximation. Nor can the levels of radiocarbon in a sample be measured accurately. Moreover, environmental factors have a significant influence on the uptake of radiocarbon by organisms. For instance, plants absorb carbon not only from the atmosphere, but also from the ground, which is largely ^{14}C-free in limestone terrain. Much the same applies to plants growing next to a busy motorway with its hydrocarbon-derived exhaust gases (e.g. Watchman 1998). Indeed, there is an intricate relationship between atmospheric $\delta^{13}C$, climate and vegetation. Different plant communities facilitate specific carbon regimes. Shrub-dominated communities of C3 plants provide values of between -12‰ and -10‰ $\delta^{13}C$ in respiratory CO_2, whereas grasses, C4 plants, produce -3‰ to +1‰ (Cole and Monger 1994). C4 plants, so called because of the four-carbon acids in which CO_2 is initially captured in their outer mesophyll cells, have a physiological advantage over C3 plants in low atmospheric CO_2 concentrations (Robinson 1994). The latter are directly related to climate, and in that sense major plant communities may themselves have an influence on climate. Numerous other factors determine the systematic errors in carbon isotope analysis. For instance vulcanism tilts the atmospheric carbon balance in favor of ^{12}C. The regime is also distorted by the de Vries effect, isotopic fractionation, exchange with the atmosphere and humic acid. The deposition of calcium and magnesium salts from aqueous solution in ground water (the 'hard water effect') provides yet another contamination source—the list seems to go on and on.

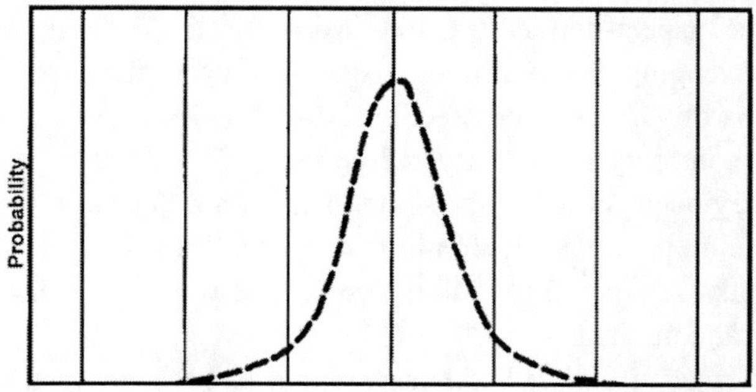

Figure 5. The probability curve for carbon isotope data being true.

Despite all these limitations of the radiocarbon method, archaeology relies heavily on it, as does rock art dating. And yet it provides no certainty, no reliability, and may produce completely false results. Contamination can also occur during sample collection or through laboratory error, for instance from oil films on supposedly sterile aluminum foil. One of the standard methods to eliminate contamination, acid-base-acid washing, has been found inadequate in removing oxalates (Hedges et al. 1998). Armitage et al. (1999) observed a 10,000-year difference in the results of two splits of a sample cleaned by two different methods. Most of the rock art age estimates published have used one of these two pretreatments, which suggests that at least some of the results are doubtful.

If this sounds skeptical, one needs to bear in mind that other supposedly reliable methods used in rock art 'dating' are probably far less credible than those derived from carbon isotope analysis. The determination of cosmogenic radiation nuclides (Davis and Schaeffer 1955), for instance, has been used (Williams-Thorpe et al. 1995; Phillips et al. 1997), providing ages that are probably random numbers. This is particularly so if the highly mobile ^{36}Cl nuclide is used and self-contradictory assumptions are made (Bednarik 1998). Similarly, cation leaching of rock varnish (Dorn 1983, 1986; Nobbs and Dorn 1988; Francis et al. 1993) is based on several logical errors and has been falsified (Watchman 1992b). Besides the inappropriate method of calibration, the use of carbon analysis of ferro-manganese accretions (which always yields false dates because of the open carbon system; Bednarik 1979), it has many procedural or logical flaws. For instance, the varnishes are highly variable, spatially (at the microscopic scale) and temporally, because they are dynamic features subject to frequent changes, e.g. through the involvement of microorganisms. The proximity of soil, oxalate, amorphous silica, or organic matter all affect cation ratios, as do lichens, fungi, ambient pH, water run-off, and surface aspect-derived relative exposure to leaching or weathering. Cation scavenging by microorganisms has been demonstrated by SEM photographs of varnish micro-sections, even by Dorn himself. Moreover, the individual solubility of the targeted cations, Ti, Ca and K, is not relevant, because they occur in minerals, sometimes within the same compounds (e.g. titanite), and it is the solubility of these minerals that is relevant. Consequently Dorn's spectral-analytical results from his samples are irrelevant random data.

Luminescence dating has been applied to rock art in a few instances, leading to a refutation in the case of the Jinmium cupules in northern Australia (Fullagar et al. 1996; cf. Roberts et al. 1998 for correction). A single optically stimulated luminescence result from a wasp nest superimposed over paint residues at a Kimberley site remains controversial (Roberts et al. 1997, 2000). It is contradicted by other age estimations (Watchman et al. 1997) and its magnitude of about 17,000 years seems to be wrong on logical grounds: it would be the world's only case of a figurative Pleistocene rock painting above ground. It is also to be questioned why the method was not combined with radiocarbon analysis, because as recently demonstrated, silicified ancient wasp nests can contain adequate organic matter to check an OSL result (Bednarik in prep.) by the vastly more reliable method. Similarly, the only attempt to use amino acid racemization (Denninger 1971) has been refuted, although it has been found that amino acid residues can be preserved in rock paints (McCarthy et al. 1994), and the method may warrant further testing.

When we consider alternative methods that have often been applied by archaeologists, the credibility of rock art age estimation work becomes even bleaker. Identification of iconography is the most frequently used approach, and yet it is largely without justification. It is non-scientific in almost every context, because identification of meaning is an unfalsifiable variable based essentially on autosuggestion. There is no evidence that past people or societies shared our subjective construct of reality, our perception or our cognition. Almost certainly these differed, and so would their determination of meaning. Very few rock art motifs are reasonably naturalistic depictions, and even then the emic meaning remains inaccessible to the alien interpreter. Similarly, the stylistic constructs of rock art 'interpreters' are entirely etic, or are 'institutional facts' (Searle 1995), or archaeofacts or 'egofacts' (Consens 2000). Individual artists may use different styles at different times or for different purposes, and the artists of a specific group do not necessarily share a common distinctive style (Mulvaney 1995; Novellino 1999). Two closely related traditional artists of the same generation might depict the same object quite differently. Imagery relating to totemic ancestors, sorcery figures, and secular meanings, produced by members sharing a cultural tradition, can be of very different styles (Layton 1991, 2001: 315). Styles as identified by non-participants of a culture are, like all archaeological taxonomies, entirely unscientific and subjective constructs of specialists (Bednarik 2011).

Similarly, there is no justification for using technique of rock art production as a credible variable to estimate age. Firstly, most archaeologists are incapable of identifying technique reliably (Bednarik 1994b); secondly, identical treatment or technique is no indication of cultural coherence, nor do differences in technique imply different cultures. Another variable often cited by archaeologists is proximity: if the rock art occurs together with occupation evidence, it might be of the same age. This argument has no logical basis, in fact the opposite could be argued just as persuasively: rock art sites are often ceremonial places, which are most unlikely to coincide with domestic sites. But more relevantly, rock art sites are invariably places of cultural focus, they as well as occupation sites are not randomly distributed across the landscape. Therefore the probability that two types of use traces, such as rock art and domestic refuse, found at one site are contemporaneous is millions of times greater at some randomly selected place in a featureless plain or desert, than it is at a site that was a focus of interest, such as a spring, cave or rockshelter (Bednarik 1989a). Archaeologists habitually ignore this rule, which leads them to propose innately improbable contentions appealing to proximity.

The most credible measures used by archaeologists to estimate rock art age are superimposition, excavation, and patination. Superimposition only provides a relative sequence, and is not always readily determined without specialist analysis. Excavation can certainly lead to credible minimum dates for rock art by one of two methods: either by determining the age of a stratum that covers in-situ rock art, or by excavating exfoliated fragments of rock art from a datable sediment layer. However, the number of published instances where this has been attempted is fairly small. The ages of in-situ petroglyphs have been estimated, at least in terms of order of magnitude, on the basis of excavation in a dozen cases (Daleau 1896; Lalanne and Breuil 1911; Lemozi 1920; David 1934; Passemard 1944; Ampoulange and Pintaud 1955; de Saint Mathurin 1975; Anati 1976: 34, 41; Rosenfeld et al. 1981; Cannon and Ricks 1986; Steinbring et al. 1987; Crivelli et al. 1996). Pictograms are very rarely preserved below ground, but they were convincingly minimum-dated through a concealing sediment at Perna 1,

Brazil (Bednarik 1989b; Pessis 1999). Exfoliated fragments of petroglyphs have been located in datable strata in six cases (Capitan et al. 1912; Hale and Tindale 1930; Mulvaney 1969: 176; Thackeray et al. 1981; Roberts et al. 1998; Bednarik et al. 2005), whereas similar claims for exfoliated fragments of pictograms are less persuasive (e.g. O'Connor 1995). Dating attempts of excavated petroglyph fragments may well provide a fairly precise timing for the event of exfoliation, but it must be appreciated that this occurred probably much later than the production of the rock art. Finally, repatination studies, if conducted in a scientific framework, can provide very useful age information for petroglyphs, as indicated elsewhere in this paper.

What follows from these considerations is that estimating rock art ages, particularly those of petroglyphs, through geoscientific means is more reliable than nearly any alternative, even though their precision may not be what some of the other methods promise. This is where we arrive at the core subject of this paper, the use of weathering and weathering-related phenomena in rock art dating.

Petroglyphs, Weathering, Accretions

As noted in the Introduction, most minerals are dissolved in aqueous solutions, but at very different rates. Rocks are generally composed of two or more minerals, which dissolve at different rates. Under ideal conditions, the determination of the time passed since the surface in question came into existence could be calculated as follows: if mineral A dissolved at $t = x$, and mineral B at $t = 10x$ then a difference in retreat of $9x$ indicates $t = 1$. This kind of approach offers more substantial possibilities for rock art age estimation than most other techniques so far employed. Similarly, weathering rinds are only suitable for order-of-magnitude age estimates, but in contrast to the mirages often conjured up by more 'sophisticated' methods, these might be nevertheless acceptable. In contrast to the often-reversible indices on which other methods rely, the processes of weathering or erosion are simply not reversible: dissolved mineral mass does not 'grow back', whereas the indices utilized in most alternative methods are either reversible, or are susceptible to variables the analyst cannot factor in. Even the use of some forms of patination very commonly associated with petroglyphs is, as recent research suggests, far more reliable than some alternative approaches used or currently contemplated. It is therefore fully justified to review these possibilities here.

Surface Retreat

Also called 'erosion retreat' or 'micro erosion', archaeologists have confused this with a very different phenomenon, microerosion (see below). Erosional retreat is the measurement of the rate of retreat of rapidly eroding rock surfaces (High and Hanna 1970; Atkinson and Smith 1976), especially those on certain sedimentary rocks, such as poorly consolidated sandstones. This retreat may be attributable to solution or granular exfoliation, and is often a combination of the two processes. The method involves the use of an engineering precision dial gauge mounted on a frame supported by three legs, and the placement on the rock of a permanent reference stud of a corrosion-free metal (Smith 1978). With this instrument, the gradual retreat of a rock surface can be monitored over a period of many years by measuring it against the reference device. The data so gathered could provide good information concerning petroglyphs at the site that have been subjected to similar surface retreat. Obviously this method would only be applicable to relatively recent rock art on such rocks as limestone and carbonate-cemented sandstone. Effectively this excludes petroglyphs that are more than a few centuries old, because on such rock types it cannot be expected that petroglyphs survive over an extended period. A good measure of the rate of retreat found in such rock types is apparent from a rock inscription the author found at a Dutch shipwreck site on the West Australian coast, occurring in a partially protected location within a rockshelter. The inscription is practically unreadable, except possibly a few letters, and the shipwreck occurred in 1712 (Bednarik et al. in press). A fake inscription at the same site has been identified through the relative retreat of the carbonate cement.

This method is therefore of limited relevance to rock art dating, and has not been used for this purpose so far. That it is likely to be of only limited applicability should not, however, discourage its further exploration. At the very least it is of considerable pertinence to issues of art conservation research.

Weathering Rind

The term 'weathering front' can have two different meanings. In the sense used by Mabbutt (1961) it denotes the contour of the deepest penetration of weathering processes into the general lithosphere. Typically this proceeds along faults and fissures and extends some hundreds of meters deep into the Earth's crust. In the sense used here, the term refers to the deepest perceptible penetration of modifications of the rock's constituent minerals within an individual clast or rock mass (Bednarik 1979). In some rock types this 'weathering zone' or 'weathering rind' is not readily detectable in the field; in others it is very clearly delineated and easily measured, being separated from the unaltered core. In most of the latter cases, the weathering rind measures in the order of millimeters or centimeters in thickness, and it differs from the unaltered rock in color, texture and other physical properties. The weathering rind of dark chert, often called flint (cf. Tarr 1938; Oakley 1939: 277; Zeuner 1960: 319; Milner 1962: 256 for corrections), is of whitish or buff color and differs from the unaltered stone in its bulk specific gravity (Bednarik 1980). The dramatic change in color manifests an etching of the microcrystalline grains' surfaces that alters the stone's reflective properties (Hurst and Kelly 1966). The character of the alteration zones of chert nodules in their natural lithological context tends to vary somewhat, which reflects the reversible nature of the replacement process that marks the diagenesis of most cherts (Walker 1962). Therefore here the light color reflects the presence of carbonates rather than the micro-pores of the weathering rind developed on individual clasts or artifacts exposed to precipitation and atmospheric buffering of pH (Bednarik 1980).

The pronounced weathering rind found on clasts or rock masses of basalts, dolerites and other rocks has been mentioned above. Trendall (1964) perceives an extremely slow development of such weathering rinds, suggesting that a veneer of 0.2 inches (5 mm) may be well in excess of one million years old. Studies of fractured stone and stone tools in the same region of coastal northwestern Australia render this figure quite reasonable (Bednarik 2007), and it also agrees with the findings of Černohouz and Solč (1966), on whose scale a weathering rind (or zone) thickness of 5 mm on basalt corresponds to 1.1 million years in central Europe. The distinctive weathering rind on mafic rocks is formed by alteration of component minerals when exposed to oxygen and water. Olivine, where present, is typically the most readily weathered mineral in mafic rocks, with pyroxene less susceptible, and plagioclase still less readily altered. Quartz, where present, is generally not affected by these weathering processes. Thus the decay of the fabric of the rock from the surface inwards first affects the iron-containing minerals, forming distinctive patterns that are visible in cross-section as rust-brown features under the microscope (Figure 6). Rainwater carries atmospheric CO_2 and other acidic anions, which are progressively neutralized as the solution percolates into the rock. The rock minerals take up hydrogen ions and water, releasing cations. Hydration and oxidation occur, and microscopic carious zones develop around individual crystals or laths. Once the process continues towards its end products, the feldspars can change to mica or montmorillonite, and eventually to kaolinite and gibbsite as the rock fabric finally breaks down. The existence of weathering rinds demonstrates, however, that the final mobilization of rock mass proceeds at a significantly slower rate than the initial etching of crystals and conversion of iron compounds; otherwise there would simply be no rind.

Figure 6. Weathering of iron-containing minerals in the weathering zone of gabbro.

The most instructive aspect of weathering rinds concerns their relationship to surface geometries: progress of the weathering front is a function of shape; it proceeds faster on projecting or convex aspects, slower on concave surfaces, as discussed below.

Accretions

Most petroglyphs have a thin cover of ferro-manganese, silica, oxalate, or carbonate accretionary deposits. The last-mentioned material was the first to be utilized in 'direct dating' of rock art (Bednarik 1981). The term 'direct dating' refers to methods based on direct physical relationship between the rock art and the dating criterion, that are applied by means of falsifiable propositions about this relationship (Bednarik 2001b: 204). This contrasts sharply with methods of inferred relationships, such as the assumption that a ^{14}C result from charcoal found in a sediment concealing rock art provides a minimum age for the art. Such propositions are constructed by chains of assumptions that cannot be falsified: e.g. that the date refers to the age of the charcoal, that the charcoal is the same age as the sediment, and that the sediment has not experienced disturbance. In direct dating, the physical relationships are unquestioned, which allows the analyst to focus on the dating criterion.

Having said this, it is also true, however, that there are great differences between the different types of accretions. Carbonate deposits occur almost exclusively as calcite, aragonite, or dolomite, may be some centimeters thick, and in some conditions may form quite rapidly. Typically they form in cavities of karst systems, precipitating from supersaturated bicarbonate solutions percolating through the rock, and include flowstone speleothems and laminar sheet deposits. In some cases, rock art has been created on such surfaces; in others it has been concealed under or even within such a series of laminae. These accretionary deposits have been subjected to age estimation by three methods: carbon isotope analysis (because ~50% of their carbon derives from the atmosphere), uranium-series analysis (such formations tend to contain small quantities of uranium), and thermoluminescence analysis (thought to detect the time the crystallization of such deposits occurred). Obviously such verifiable results are from substances that are physically and indisputably related to rock art, being either older or younger than it; they offer direct dating evidence (Bednarik 1980, 1984, 1999).

Silica skins (discussed above) can contain bacteria, pollen, fungi, or algae, which have already yielded a radiocarbon date from a silicified wasp nest of the Pleistocene (Bednarik in prep.), and also numerous mineral substances such as clays. Ferro-manganese skins may comprise both accretionary and endogenous components, the latter deriving from weathered primary ferro-manganese minerals in the host rock. However, it is suggested that the exogenous component is significantly greater, and that the portion defined as inherited weathering products (the product of ancient deep weathering) is minimal (*contra* Donaldson 2011) and has in any case been recycled. Iron and manganese play an important role in the metabolism of lithotropic bacteria (e.g. *Ferrobacillus, Leptothrix, Spirophyllum, Thyobacillus, Gallionella*), deriving energy from the oxidation of bivalent iron and manganese (Bazylinski and Frankel 2003; Kappler et al. 1997; Konhauser 1997; Perez-Gonzalez et al. 2010). These and anoxygenic photoautotrophic bacteria have catalyzed the principal cations of the ubiquitous ferro-manganese deposits almost since life first appeared on the planet and their recycling has continued ever since. Therefore most components of such an accretion today are not of an age corresponding to its antiquity, nor can a bulk sample of its carbon component be expected to yield an age resembling the time of its deposition. Yet the method of bulk-sampling such accretions for radiocarbon analysis (and even for calibrating another shortsighted dating method, cation-ratio determination) has been the most-used so far in attempts to determine the ages of petroglyphs. This has been the case despite this author's warnings for over 30 years that the carbon system of these features is open, and that this method can therefore only produce false datings.

On the other hand, this does not absolve ferro-manganese accretions from becoming involved in the estimation of petroglyph ages. But the methods to be engaged in this quest will need to be quite different than those applied to them in the late 20th century, and one, via colorimetry, has been tried and has yielded surprisingly consistent and credible results (Bednarik 2009b). Although these mineral accretions can only survive at neutral to alkaline conditions at the atmosphere-lithosphere interface, and are being depleted by the widespread atmospheric acidification experienced by the Earth, they still remain very common and occur together with the majority of surviving petroglyphs. They are found in different forms, ranging from the dense rock varnishes predominantly of arid regions, to less distinctive forms of essentially *dunkle Rinden*.

Nevertheless, colorimetry and analyses of accretions occurring with rock art—whilst providing securely relatable empirical evidence in rock art age estimation—provide only experimental results. Those likely to be secured from actual weathering phenomena may lack precision, but provided their significance and limitations are fully appreciated, they are certainly the most reliable. But before developing a specialized methodology it would be useful to explore some of the inherent principles of the processes that lead to the reduction of rock to sediment. The various mechanical and chemical factors involved are being considered in other contributions to this volume. Here, a quite specific aspect of rock weathering will be considered in an attempt to derive some 'first principles' of rock weathering as it might pertain to rock art study and to archaeology.

The Geometry of Rock Weathering

The Geometry of Stone Tool Taphonomy

While stone artifacts are not rock art, they are often used in archaeological arguments about rock art, and the geometry of their weathering is a convenient starting point for the generic subject of the geometry of rock weathering. Most stone tools are made from sedimentary silicas, usually cherts, including flint (Tarr 1938; Oakley 1939: 277). Cherts are fine-grained silica rocks comprising mostly silica (SiO_2) and calcium carbonate ($CaCO_3$) and occurring most frequently in limestone facies. The practice of defining nodular microcrystalline and chalcedonic quartzes as flint may reflect their mode of occurrence but is geologically invalid (Zeuner 1960: 319; Milner 1962: 256) unless they are indistinguishable from flint in every important physical aspect. Since Judd's (1887) attempt to explain the 'patina' of flint and other cherts there has been much speculation about this form of weathering rind (Tarr 1926; Correns and Nagelschmidt 1933; Gehrke 1935; Donnay 1936; Bosazza 1937). As the correct hints by Washburn and Navias (1922a, 1922b) remained unheeded until 1951, archaeological writers of the time (e.g. Curwen 1940; Mitchell 1947) had no choice but to subscribe to the opaline content theory. However, since the work of Midgley (1951) and Weymouth and Williamson (1951), and the overwhelming evidence provided by Folk and Weaver (1952), there remained no justification for repeating earlier fallacies.

Diagenetically, cherts are the result of selective replacement of calcium carbonate by silica at normal ambient temperature, forming as nodular deposits that may coalesce into tabular formations following the limestone host rock's strata. As each crystal is gradually replaced, the chert's fabric copies that of the former limestone faithfully in every detail, so that fossil casts in it are precisely replicated. The nodules of chert are surrounded by zones of partial replacement, often called their 'cortex' (Figure 7), and since the process of replacement is reversible (Walker 1962), subject to environmental fluctuations, there may be concentric formations present that indicate variations in silica content (Bednarik 1980). The darkest zones are those richest in silica, culminating in black flint. The color is simply a function of light reflection: a fully silicified stone absorbs all light and is therefore black.

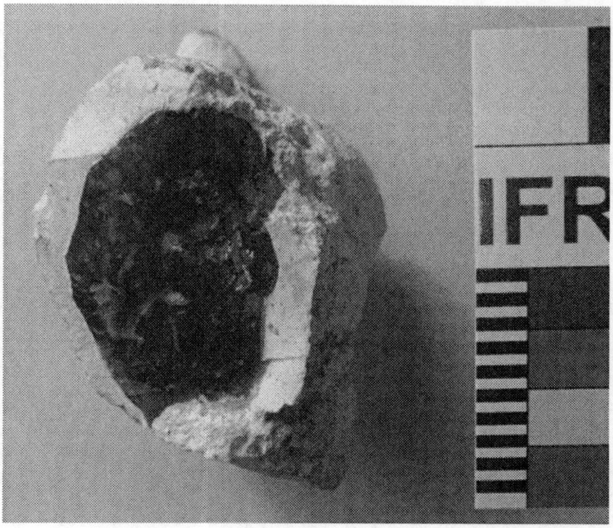

Figure 7. Sectioned flint-like chert nodule from Mt Gambier, South Australia, with white cortex.

The light-colored weathering rind of chert, by contrast, is the result of mobilization of silica in the form of monosilicic acid (Mosebach 1955). Schmalz' (1960) view on flint patination contrasted with that of Oakley (1939), who considered carbonic acid in the ionized state to be the chief patinating agent. Schmalz, however, misunderstood the crucial passage in Weymouth and Williamson (1951) and since he was apparently unaware of both Midgley (1951) and Folk and Weaver (1952), he believed to be the first to demonstrate the absence of opal in flint. Clark (1979) noted the diverging concepts of some writers on the role of calcareous sand. Mitchell suggested repeatedly (e.g. 1947: 299) that carbonate waters leached silica from the surface of cherts, but his assumption that the specific gravity of flint is higher than that of the cortex contradicts his espousal of the interstitial opal theory (removal of colloid silica would increase, not decrease, the S.G., but would decrease the bulk density G_S). In that he followed Moore and Maynard (1929: 276) who had defined them as the most effective solvent of silica from norite and diabase. Curwen's (1940) experiments indicate the involvement of both organic and calcareous matter, bringing to mind an early comment (Murray and Irvine 1891) that siliceous remains are appreciably soluble in cold seawater in the presence of $CaCO_3$ and decaying matter (cf. Fairbridge 1964: 444–7). The tenability of the implied process is, however, challenged by the tendency of many organic compounds to interact with carbonate mineral surfaces, forming mono- or multi-molecular layers isolating the carbonate from water and inhibiting even inorganic equilibrium (Suess 1970).

All early commentators believed colloid silica to be the only form present in natural waters. Because it was proposed only in the year of their own publication, Kahlenberg and Lincoln (1898) lacked the benefit of the colorimetric silica determination method, but the several later writers repeating their view (e.g. Tarr 1917: 433, 1926: 25; Twenhofel 1932: 44, 510; Van Tuyl 1918: 456) did not. Diénert and Wandenbulcke (1923), enlarging upon Jolles and Neurath's (1898) method, were unable to detect any colloid silica in their samples of river water and in the following year confirmed its absence by their failure to raise the molecular silicate content of water by increasing its pH, which should convert any colloid fraction present (Diénert and Wandenbulcke 1924). Harman (1926) scrutinized Kahlenberg and Lincoln's (1898) propositions and concluded that they are untenable in the light of his meticulous and varied experiments. After pursuing established concepts of another discipline, Roy (1945) asserted that silica of natural waters is molecularly and not colloidally dispersed. Aoki (1951) suggested molecular dissolving of silica in water up to a certain concentration, claiming that surplus silica (which would be produced by a lowering of temperature or pH) will polymerize to the colloid state (cf. Tsai et al. 2006). Whilst solubility in the range of pH 0–9 differs little (Krauskopf 1956) and may in fact be slightly greater in acid than in weakly alkaline conditions (White et al. 1956)—at least at low temperatures (Okamoto et al. 1957: Figure 2)—it increases dramatically above pH 9, where silicic acid only polymerizes in high concentrations (ibid.: 126). Silicic acid will of course dissolve in water to the concentration of equilibrium solubility, which also rises rapidly above pH 9. Iler (1955: 25) reported that in highly alkaline environments the ion $H_3SiO_4^-$ becomes important to the point of dominating over H_4SiO_4 above pH 10, and it is generally acknowledged that ionic forms are present in appreciable amounts only at >pH 9. Thus formation of a weathering rind on microcrystalline sedimentary silica is attributable to the etching of the microcrystalline grains' surfaces, which alters the reflective properties of the stone (Hurst and Kelly 1966).

Since the process can occur on the surface of soil sediment and in the absence of limestone, but is almost ineffective in sheltered conditions even after burial for hundreds of millennia, frequent exposure to moisture under ambient conditions of high pH is the probable cause. Both the cortex and the weathering rind ('patina') of cherts can absorb iron salts into its pores and acquire a brownish color (Bednarik 1980). The weathering zone tends to be better delineated than the cortex, and the fact that it exists demonstrates that the progress of solution at the weathering front must be significantly faster than the total reduction of the surface crystals (the surface retreat rate). This is despite the immensely increased surface area within the altered zone and the outer zone's higher penetrability to alkali replenishment, which would not be so readily attainable at the weathering front, where the silicic diluents would be both nearer equilibrium and less readily transferable. For instance, if a 1-mm-thick weathering zone is immersed in sodium hydroxide solution (saturated at its boiling point), exposure for one hour will result in its almost complete removal, without corrosion of the structurally unmodified core (Bednarik 1980). Therefore the main factor in the progress of the weathering is the frequent flushing of the interstitial spaces of the weathering zone. This is confirmed by the observation that the upper surface of specimens exposed to precipitation exhibits a thicker weathering zone than the underside.

The geometry of the weathering process is here of particular interest because it introduces the subject of the way weathering generally proceeds. Attempts to gain some insight into the factors influencing patination rates of clasts or stone tools by quantifying their effects identify weathering progress either by patina thickness (P_t, cf. Figure 8), or by total weight loss through weathering (see Legend of Symbols):

Legend of Symbols:
V = Bulk volume
G_S = Total bulk *S.G.*, or *W/V*
L = Maximum length
E = *Größte Dicke*, after Cailleux (1951), or $\pi_L \times L/100$
P = Percentage of patina (by volume)
π_L = *Abplattungswert*, after Lüttig (1956)
A_i = *Abplattungsindex*, after Cailleux (1951)
P_t = Patina thickness
α = Patination index; or $S.G._{silicate} - G_{S\ (patina)}$
β = Porosity index, % water accepted after 90 seconds (by volume)
S.G. = Specific gravity
W = Weight, after drying at 105°C until no further loss
W_S = Weight, immersed in water at 20°C, waterproof coating discounted and specimen fully dried
W_{90} = Weight of dried specimen in water at 20°C, after immersion for 90 seconds

W_L = Total weight loss through patination

$$W_L \text{ (in \%)} = \frac{(2.55\ V - W)\ 100}{2.55\ V} \tag{2}$$

Alternatively, percentage of weathering by volume can be calculated:

$$P = \frac{255 - 100\ G_S}{\alpha} \tag{3}$$

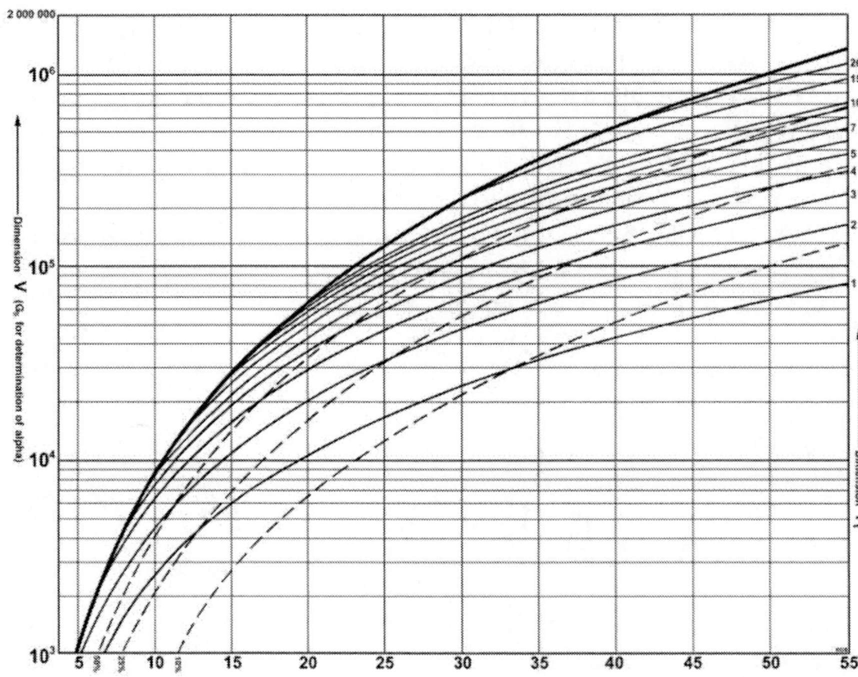

Figure 8. Relationship of G_S, P, α and P_t, for cuboid of π = 25 (being a function of A_i = 3).

To eliminate the need of physically measuring P_t (which usually involves damaging the specimen), α can be estimated by an experienced observer (it can only vary from 0.5 to 0.9) and theoretical P_t can be determined by translating the principles established in Figure 8 into seven coordinates. In practice, a difficulty to be taken into account is that posed by the possible presence of 'gray zones', which can introduce an unknown variable. They are patches of gray within the weathering zone, presumably where silica crystallization had resisted weathering locally. Such 'gray zones' are present in early specimens possessing a G_S >2.2, except those of W >100 g. In one sample analyzed the *S.G.* of 'gray zones' ranged from 1.95 to 2.20 (Bednarik 1980).

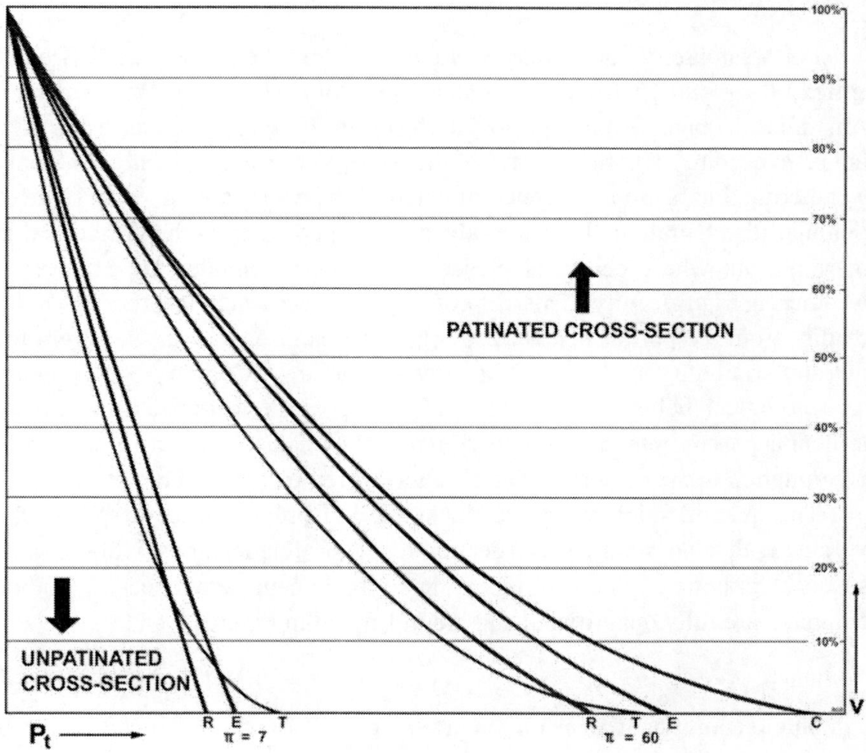

Figure 9. Influence of cross-section shape and Index of Flatness π on patination progress: C = circular, E = elliptical, R = rectangular, T = triangular. Two Indexes of Flatness (7 and 60) are considered.

As readily demonstrated by Figure 9, the Index of Flatness (depending on application, either that of Cailleux [1951] or of Lüttig [1956] was employed) is of far more consequence than implement shape, as expressed by nominal cross-section (which was divided arbitrarily into circular, elliptical, segmentary, rectangular or triangular, to show the great difference in rate of weathering relative to shape), when relating P_t to cross-section. To assist with the estimation of α, the porosity index β was established. Porosity or void content can be ascertained either with the McLeod porosimeter or with the wax-coating method.

$$\beta = \frac{(W_{90} - W_S)\,100}{V} \tag{4}$$

For a sample of early Holocene and late Pleistocene chert tools (free of cortex, W = <80 g) from 39 coastal sites along Discovery Bay, southern Australia, β ranges from 1.92 to 20.31 (mean 7.36). In general, a porosity index exceeding 10, for flakes of the sizes surveyed, indicates not only weathering, but also the presence of extensive leaching and an early stage of granular disintegration. The latter always commences from the artifact's thin margins, but where cortex is present this, too, accelerates the process of weathering significantly. This is probably because leaching proceeds more readily from the porous contact zone (larger area of exposure) than from the initially smooth surface of a fresh fracture. Contrary to popular archaeological belief, cherts can totally disintegrate within a few tens of millennia in environments found commonly in nature, such as exposure to precipitation in the presence of an alkaline buffer, e.g. at sea coasts.

The quantification of patina thickness and proportion of a specimen's volume is thus possible without sectioning. Their relationships with total and specific gravities, flatness index and implement cross-section, albeit complex, are fully quantifiable and are illustrated in Figures 8 and 9.

The Geometry of Erosion Weathering

Having thus opened an investigation into the geometry of rock weathering, the possibility of determining the age of rock surfaces has been considered by archaeological investigators for at least two centuries (e.g. Belzoni 1820: 360–361). Two Czech geologists, J. Černohouz and I. Solč (1966), tried to solve part of the problem and described a method of determining the ages of blunted edges on sandstone and basalt. They used the angle of the edge, the distance of retreat at the edge, and two 'experimentally found' constants (a, b). The age of the edge t can then be determined from the distance of its retreat h and its angle φ:

$$h = \frac{a\ t\ (2)}{1 + b\ (t)} cot \frac{\varphi}{2} \qquad (5)$$

The age was reportedly determined with an accuracy of ~10–20%. However, the distance of retreat from the originally sharp edge cannot be measured (Figure 10), because the rock surface recedes not only from the edge but also from the two surfaces forming it (Bednarik 1979). Therefore the original dimension *h*, at the time erosion commenced, cannot be measured (see Figure 11, which shows that the actual edge retreat is not *h*, but the sum of *h* + *x*, *x* being a function of *z*). Although the theory has thus been refuted, the underlying principle is sound, and such rounded edges do relate to their ages. However, the solution to the quantification of the process was not available until a quarter of a century after Černohouz and Solč's proposal (Bednarik 1992).

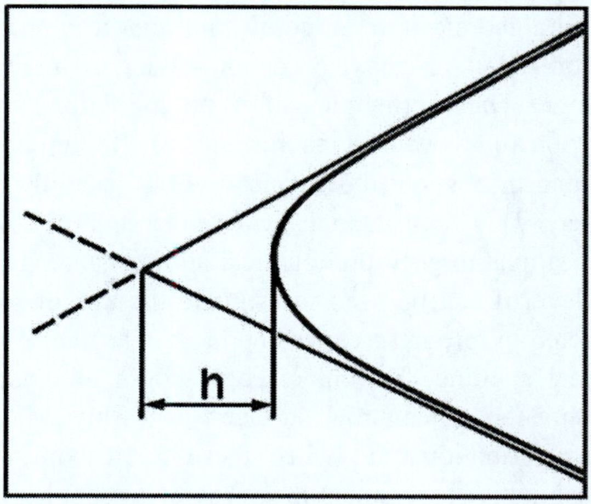

Figure 10. Development of sandstone wanes as assumed by Černohouz and Solč (1966).

The 'radius' of rock wanes (in geological usage, 'wane' refers to the progressive rounding of a rock edge) clearly does increase as a function of age. It should be noted, however, that the retreat at the edge 'does not occur as the simple curvature shown by Ollier (1969: Figure 149), it consists of a hyperbola in section' (Bednarik 1979: 28). Although wanes are thus not equicircular in section, to simplify the following explanation it will be pretended that wanes have a radius; the resultant error is not significant to the overall issue. It is of greater consequence to appreciate that the process must be seen as an attempt of an eroding mass to attain a state approximating equilibrium (i.e. where erosion coefficients would affect the surface uniformly). Inevitably, this leads to a spherical form in the case of shape. The process thus conforms to geometrical axioms, in precisely the same way as heat is transferred proportionally within a solid body, or for that matter, the pattern of penetration of a cube of sugar dropped into coffee is that it occurs faster from the corners than on the flat sides; and the same applies to a melting solid. This pattern follows a universal law, according to which the unaffected core increasingly resembles a sphere (Bednarik 1992). Or, geometrically speaking, the unaffected zone trends towards the smallest surface area relative to mass, achieving that perfect shape at its vanishing point.

This means that heat transfer in a solid object, or weathering of a clast, or melting of a piece of ice, must proceed faster from convex or protruding surface aspects, and relatively slower from concave aspects. The process has been defined in algebraic fashion, confirming that it is quantifiable and that wane formation, be it on rock, or on dissolving or melting objects, is a function of time. These principles of wane formation are universal, they apply microscopically as well as macroscopically, to individual crystals or to rock masses, and throughout the physical world generally (e.g. relative rate of heat transfer). This fundamental law (Bednarik 1992) offers a variety of practical applications in geomorphological and archaeological dating. At the macroscopic level it can be used to estimate the age of rock edges, where these are affected by progressive weathering rather than by episodic events, such as thermal spalling. The microscopic edges of crystals fractured, or truncated by abrasion, when a rock surface was freshly exposed are subjected to micro-wane formation, and when these microscopic wanes are clear enough for quantitative assessment they provide a reliable measure of the age of the rock surface in question (see next chapter).

The theoretical principle of quantification is depicted in Figure 11. In wane formation, the ratio $h : r$ is constant for any angle α, irrespective of distance of retreat of the faces and the edge. Dimensions h, y, and A can be measured, and r can be either determined physically (but only approximately, being a hypothetical value) or it can be determined from α and y:

$$r = y \tan \frac{\alpha}{2} \qquad (6)$$

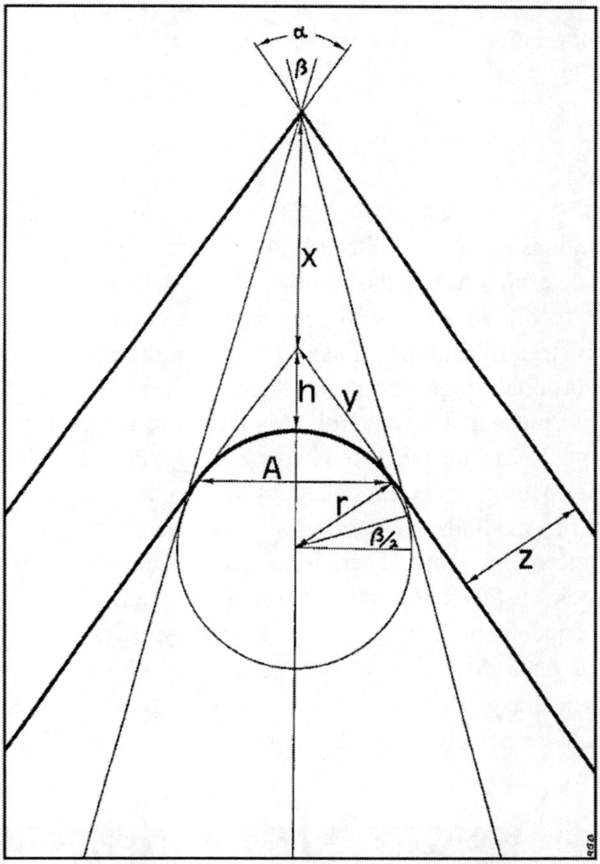

Figure 11. Diagram depicting the universal laws of wane formation in simplified fashion.

Ratio $x : z$ is a function of α, and at α = 60, $x = 2z$. The ratio $h : r$ is constant for any angle α, irrespective of the amount of retreat. Dimension x can be expressed in algebraic fashion:

$$x = \sqrt{\left[\left(\frac{z}{\tan 0.5\alpha}\right)^2 + z^2\right]} \tag{7}$$

This leads to the prediction of β, the angle expressing the rate of wane development relative to surface retreat:

$$\beta = 2\sin^{-1}\left(\frac{r}{x+h+r}\right) \quad (8)$$

The relationship wane width A with age, irrespective of actual retreat, is ultimately determined by the ratio $\alpha : \beta$, which must be established empirically. It follows that the dimensions A, r, z, and angles α and β are all related geometrically and algebraically, and that the variables A, r, x, z and h are all proportionally equivalent, and increase linearly with age. Of these, A is most easily measured physically. It is therefore the variable used in wane measurement. It can be measured with a gauge bearing a full selection of semicircular openings of calibrated sizes held to the rock wane. Despite both the ease of the method and its obvious potential, it remains little used. Of particular relevance is the observation that the development of weathering rinds on rock surfaces follows the same geometrical laws: on any rock surface of equal exposure it will be markedly thicker on any prominent aspects than on a flat surface, and conversely it will be thinner beneath concave surface aspects (Figure 12). Therefore it can be assumed that here, too, the unaltered core would approach spherical shape before its disappearance.

Figure 12. Section through weathered dolerite clast, showing deeper penetration at corners and shallow penetration under concave surfaces.

The Geometry of Microerosion

Instead of developing the measurement of rock wanes as a tool of estimating weathering duration, the geometric principles of wane development have been more often applied at the microscopic level. As a purely optical technique, microerosion analysis was the first direct dating method for petroglyphs that involves no destructive removal of samples, is of very little cost, and provides statistically meaningful data; under favorable conditions, hundreds of measurements can be taken from a single sample. It is also expected to be more accurate than other methods of dating petroglyphs, especially once it has been tested in different geological and climatic environments (Bednarik 1993a). Obviously it has a variety of potential applications in archaeology, such as the determination of approximate age of stone structures, quarries, monuments, statues, and even some stone tools. However, its intended main use is in petroglyph dating, and it was conceived primarily to provide a relatively simple method of reliable age estimation of geomorphic exposures (Bednarik 1992, 1993b, 1995, 1997, 2000b, 2001b: 129ff, 2002a, 2010; Tang and Gao 2004). The method was initially based on the very same geometrical principles defined in the previous chapter, and applied to the microscopic wanes forming on fracture edges of particularly erosion-resistant minerals, especially quartz.

This method is has now been applied in several blind tests, having been used to estimate the ages of petroglyphs in all continents, in numerous countries. In all blind tests it has met prior archaeological predictions. Where surfaces of known ages (such as those of glacial striae, graffiti, gravestones, monuments, dated inscriptions, stone bridges, stone sculptures, etc.) are present at a site, a calibration curve can sometimes be derived from their micro-wanes, leading to the dating of a surface of unknown age (e.g. a petroglyph) in the vicinity. To limit the possible effects of past climatic fluctuations, the microerosion coefficients of two or more component minerals of composite rocks are compared from the same surface, such as those of quartz and feldspar on a granite surface. The method is not eligible for rock surfaces that were covered for any great length of time, be it by soil, rock varnish, water or whatever else; nor can microerosion dating be used where a rock surface has been subjected to the application of certain chemicals or has been physically worn, for instance by the activities of graffiti removers.

In considering microerosion dating we have to distinguish between two types of rock: those on which the individual grains or crystals that were fractured, exposed or truncated during the manufacture of a petroglyph are capable of remaining in situ for periods exceeding the age of the petroglyph tradition in question; and those sedimentary rocks that are subjected to relatively swift granular exfoliation or chemical corrosion, such as carbonate-cemented sandstones, calcite, dolomite, etc. (cf. Emery 1960). Only the former are of interest here, because for microerosion analysis to be possible, the presence of crystals or grains truncated (through impact fracture or abrasion) by the rock artist is absolutely essential. Moreover, certain rock types are inherently much better suited than others, and at least initially composite rocks which include quartz (or another comparatively stable component), such as granite, quartz porphyry, gneiss, diorite, granophyre and the like, are more likely to produce valid results. As reliable quantitative information becomes available, other rock types will become susceptible to the method.

The rates of micro-wane development may vary somewhat through time; they are likely to be susceptible to environmental factors, but such fluctuations may cancel each other out over long time spans. However, of the three main variables, two have no effect on the relatively coarse results: ambient pH and temperature have hardly any impact within the range of natural conditions, and moisture availability can be accounted for by checking the response against a second mineral that is differently affected. Obviously there are also variations to be expected according to relative location of an edge (i.e. relative to prevailing climatic factors, exposure to anomalous influences such as abrasive wear, and other factors), but these have no bearing on the regularity of the underlying theoretical principles of the wane-forming processes as described. These factors are the method's only potential major error source, but while they can influence its accuracy they can certainly not affect its intrinsic validity. The only critical comments made so far (Pope 2000) raised irrelevant objections, e.g. that crystalline quartz occurs in two different forms. That is correct, but beta quartz is only stable at temperatures above 1063°F (573°C). Thus, all quartz in nature is alpha quartz, because once a sample of beta quartz is lowered below the mentioned temperature, it automatically transforms into alpha quartz.

To limit the possible effects of past climatic fluctuations on the results of microerosion analysis it us useful to compare the microerosion coefficients of two or more component minerals of composite rocks. In the example from Lake Onega, Russia, where the method was first applied (Bednarik 1992, 1993b), those of quartz and feldspar were compared. It is reasoned that, whatever the past climatic fluctuations were, and however they affected the minerals, it is likely that two different minerals would have been differently affected, and this would be reflected in differences in the two calibration curves. If the related values on the two calibration curves correspond on the ordinates it suggests that erosion rates have not been sufficiently affected to distort age estimates significantly, and the reliability of the result is considerably enhanced.

While this method was conceived initially for estimating the age of petroglyphs, it is of course valid for any geomorphic surface. It can be applied to any rock surface where edges of crystals were freshly exposed to weathering, provided there are at least remnants of the crystals' surfaces still present for examination. However, on present indications it would appear that the susceptibility of quartz to the method would be limited to a maximum of a few tens of thousands of years, perhaps 50,000 years at most, because beyond such age the effects of microerosion would not permit reliable measurement.

The immediate, tangible advantages of the microerosion dating method are its most impressive aspects:

- It is cheap and reliable, and requires little training apart from field microscopy.
- It offers very considerable scope for development, being in fact a variety of methods, most of which have hardly been explored.
- It is the only currently available petroglyph dating method that seeks to date the event of petroglyph manufacture itself (the 'target event' of Dunnell and Redhead 1988) rather than a physically related feature that merely provides minimum or maximum age.
- It is the only viable rock art dating technique currently available that involves no sample removal.
- It involves no possibility of sample contamination, as practically all other dating methods do.
- It offers a mechanism of internal crosschecking when two or more component minerals are considered.
- It is the only currently available rock art dating method that has the capacity of satisfying statistical sampling requirements, because it can furnish vast numbers of age-related values from a single motif.
-

Perhaps most importantly, microerosion dating will induce researchers to scan petroglyphs very carefully in the field, prompting the finding of much forensic evidence concerning the history of the motif and its support surface that would otherwise be overlooked: later modifications to the motif, use traces, traces of aeolian wear, fire damage, glacial or sediment scouring, frost or regelation damage, deposits of precipitates and their alterations through time, patination and weathering, effects of vandalistic recording practices, etc. Indeed, where microerosion dating differs most from previous methods of rock art age estimation is that the latter are rather like hunting expeditions, from which researchers bring their samples to the 'safety' of the laboratory. Microerosion analysis involves lengthy field studies, but no laboratory work at all (the results can be determined in the field). Finally, it reverses the underlying strategy of previous attempts to date petroglyphs directly: instead of using non-archaeological data to produce minimum or maximum dates of the art, it utilizes surfaces of known ages to calibrate a non-archaeological process (erosion), which is then used to provide a time frame in which to place petroglyphs of unknown age.

Finally, alternative methods of microerosion analysis are possible, one of which has already been applied occasionally. This is the examination and measurement of alveolar development on the floor of petroglyph grooves, where more erosion-resistant minerals project above the more soluble components (Bednarik 1995: Figure 10). The relative relief differences are a function of time, becoming greater with increasing age, but no attempt has been made so far to quantify the process.

The Geometry of Sandstone Shelters

When we look at the world's quantified distribution of surviving rock art, it is immediately obvious that most of the major rock painting occurrences are found in four regions: southern Africa, India, northern Australia, and the Noreste of Brazil. Petroglyphs dominate numerically in the rest of the world. A logical explanation is that during the Mesozoic these areas were joined as the supercontinent Gondwana, and they have similar extensive sandstone sequences. This is fairly obvious to the traveler, through similar landforms, hill shapes, erosion patterns, hydrologies and even vegetation regimes. Most relevantly in the present context, these sandstones are more susceptible to rockshelter formation than practically any other rock type. Rock paintings, wherever they may have been made, can only survive in sheltered locations for any prolonged period of time. A small number worldwide has been preserved in deep limestone caves; the vast majority of pictograms occur in sandstone shelters. In addition, weathered sandstone surfaces tend to be pervious to some degree, allowing mineral pigments to penetrate slightly, which adds to their ability to survive, or provides the potential of traces to persist after the paint residues have largely eroded.

But how do these rockshelters form in sandstone, especially in specific types of this rock? There are essentially two processes at work: granular and laminar mass-exfoliation. Both are largely (but certainly not entirely) caused by the same factor, capillary moisture in the rock, which rises from the base and is buffered by the contiguous sediment deposit. Every time the moisture rises, it weakens the rock's fabric by mobilizing solubles, leading to the fretting of small particles or individual grains. But the process also leads to the formation of subcutaneous salt deposits (subflorescence), usually in a zone 10–20 mm from the surface and parallel to it. In due course *Salzsprengung* (fracture through expansion of salts, such as gypsum, anhydrite, or a chloride or nitrate) causes the exfoliation of a laminar rock flake. Since this process is more effective near the floor (i.e. within the zone of effectiveness of rising capillary moisture) and tends to be ineffective above a certain height from the floor (as a function of duration and severity of the moisture regime), it logically has to result in the formation of a concavely shaped rear shelter wall. Over the course of centuries and millennia the succession of countless rises of interstitial moisture, which are most effective just above the floor, results in a more or less curved profile, with the greatest shelter depth typically at floor level.

Obviously the depth of these sandstone shelters is in some fashion a function of their age as this exfoliation process gradually forms them. However, if one examines a large enough number of excavated shelters a curious observation can be made. Below the sediment floor, the wall does not recede further, as it would due to a rising sediment level. On the contrary, it tends to slope towards the shelter entrance (Figure 13). If the deposit is deep enough one can even reach a depth at which it reaches the projected drip-line, where the rock may continue vertically.

This phenomenon is explained here as follows: exfoliation, especially the laminar kind, is ineffective below floor level, because the rapid wetting and drying (or freezing and relegation) cycles are not effective within the sediment cover. Moisture and temperature levels are much less variable, and insolation is absent. If the exfoliation is most effective just above the floor, and if the floor of the sediment is gradually rising (as may be the case in most shelters), it follows that the forward-sloping rock ledge within the sediment is the result of the rising sediment. If fresh exfoliation is limited to the wall above the floor, and if the floor deposit rises, it follows that as the depth of the shelter increases, there should be a sloping rock shelf below the sediment, under ideal conditions.

Moreover, this geometry also shows that the inclination of the sloping rock must be a predictable function of sedimentation rate, exfoliation rate and time: the steeper the slope, the faster sedimentation and the slower exfoliation would have proceeded. Under perfect conditions, the three variables would be proportionately related, but such conditions might be rare. Both sedimentation and exfoliation rates are usually variable through time and obviously depend on various other factors. However, as a theoretical principle determining the geometry of rockshelter sections this is a useful rough guide for the researcher. It again indicates the utility of applying geometric principles to issues of weathering.

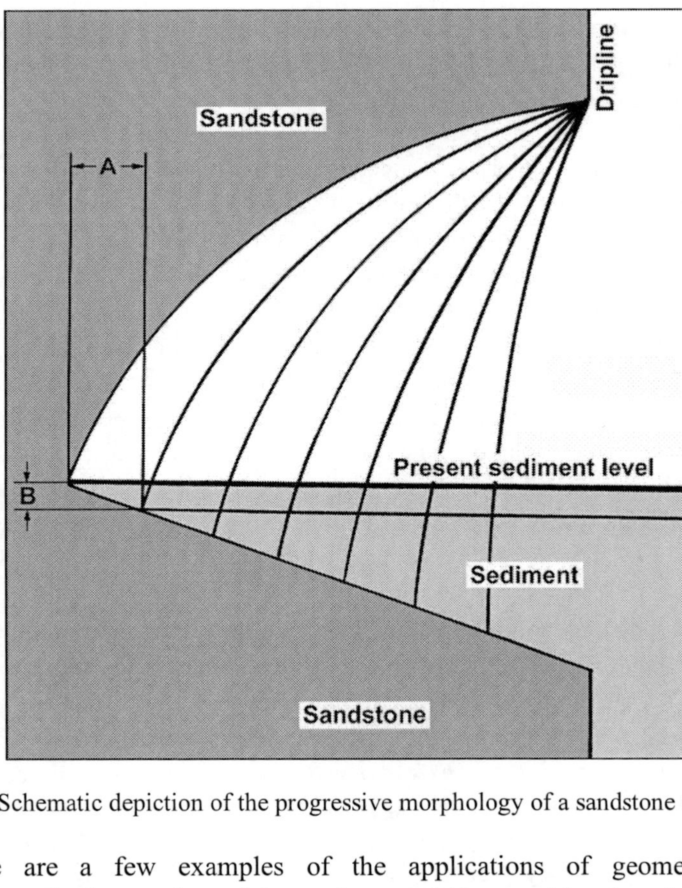

Figure 13. Schematic depiction of the progressive morphology of a sandstone shelter.

These are a few examples of the applications of geometry in assessing weathering products. Many others could be considered, such as, for instance, the laws determining the formation of solution scallops in limestone caves, with their certainly distinctive geometric properties (de Serres 1835: 24; Bretz 1942; Curl 1966, 1974; Shaw 1992; Mihevc et al. 2004: 522)—provided the term 'weathering' is extended to processes of speleomorphology. But what these examples show is that once the laws governing the weathering processes are understood and rendered quantifiable, they permit the determination of the relevant variables, most especially time. In that sense many of such phenomena can be utilized in estimating the ages of geomorphic exposures, such as petroglyphs. That potential remains under-explored and under-utilized, perhaps because a dependency has been allowed to develop, especially in rock art dating, on over-sophisticated techniques that promise more than they can deliver (Bednarik 2002b). In considering the severe limitations of these alternative techniques, the complex interdependencies, supervenience relationships and qualifications applying to them, it soon becomes evident that, at least in petroglyph dating, geomorphological methods are more reliable. While it is apparent that their degree of precision may leave much to be desired, it is perhaps the case that reliability is preferable to a promise of precision that may not be deliverable. Rock weathering and related features, such as the fluvial erasure of inscriptions or petroglyphs in valleys subjected to inundation (governed by the geometrical laws of the behavior of suspended sediments; Bednarik 2009a), offer the most reliable basis for quantifying changes to geomorphic exposures. That potential has not been adequately explored so far, and for this to occur it is essential that the underlying laws, especially the geometry of these processes, be defined.

Weathering and Rock Art Conservation

The weathering of rocks is important in the field of conservation and preservation as well, not only in the determination of the age of petroglyphs. There are three basic processes of weathering with relevance to the development of appropriate strategies for cultural resource management practices: chemical, physical, and biological.

Chemical Weathering

The most important factor in chemical weathering factors is the presence of water and its dissolved constituents. Water occurs at rock art sites in various forms:

- as phreatic reservoirs (exposed aquifers, especially in limestone caves),
- moving or stationery surface deposits (lakes, rivers, waterholes),
- concealed aquifers,
- surface run-off,
- atmospheric precipitation,
- gravitational water percolating through a rock mass,
- capillary moisture rising in a porous rock mass,
- condensation surface deposit, bound within the crystal structure of component minerals,
- interstitial moisture in equilibrium with relative atmospheric humidity,
- and in the form of air humidity.

The interdependence of many of these forms of moisture (e.g. capillary water requires a source such as an aquifer or episodic supply) is crucial in understanding the hydrological regime of a site, which in turn is essential in any remedial work.

There are several deleterious effects caused by most forms of water occurring at a site, while at the same time the presence of *some* moisture may be essential for preservation. Water traveling within the rock, be it gravitational or capillary, generally acquires solutes (typically chlorides and carbonates), especially because the pressure within a rock mass is significantly greater than atmospheric pressure (it can be hundreds of times greater). Pressure is one of the most important variables determining solubility, the others being temperature, kinetic turbulence, pH and presence of corrosive compounds. This means that when such vadose water re-emerges at a surface, where its pressure suddenly reverts to atmospheric conditions, most of its solute has to be precipitated. Stalactites, for instance, are formed by this process, as are numerous other forms of deposits found at rock art sites.

Weathering zones on rock are, as noted above, primarily the result of solution, hydration, hydrolysis, ion exchange, chelation, oxidation and reduction, leading to the selective removal of mineral mass. Moisture is thus heavily implicated in rock weathering, but as a cause of weathering zones it is usually most effective as direct precipitation. There are exceptions, however, particularly in the case of sandstones, where granular exfoliation of sheltered areas can greatly exceed the effects of weathering processes on surfaces fully exposed to precipitation. Only modest quantities of moisture are involved here, which are sufficient to gradually dissolve the carbonate or silica cement, and which are assisted by the formation of salts that wedge individual grains loose as they expand. As already noted, the resulting granular or mass exfoliation affects vast numbers of rock art sites.

Condensation water forms in caves and rockshelters when the relative air humidity is high and the ambient air temperature is greater than the rock face temperature. The dew point is the temperature at which air becomes saturated and dew is precipitated. Limestone caves frequently have a relative air humidity of 95% to 99%, yet condensation is not necessarily common in caves, particularly those where the speleoclimate is almost entirely independent of the atmospheric climate. Nor is the water, which condenses in distilled form, capable of dissolving the limestone in the absence of carbon dioxide. In a deep cave system, air temperature is buffered by rock temperature and therefore very stable. There is practically no convective temperature exchange in caves of a high air-residence rate, and the temperature differences necessary to induce condensation do not exist. Condensation is more likely to appear near the entrance of a cave, or in rockshelters. It forms as water beads or as a film that can cause significant damage to paint residues (paints were often based on water as the solvent). It also tends to attract cryptogamic growth within the reach of light (artificial or natural), and it can deposit a variety of dissolved salts. In the case of carbonate rock, such surface water deposits can even form carbonic acid in the presence of carbon dioxide, and convert the rock to water-soluble and mobile bicarbonate.

Surface run-off is another factor frequently implicated in the destruction of rock art (Figure 14). On pictograms it can be highly effective, removing unconsolidated pigments rapidly. Where it occurs regularly, the establishment of mosses, fungi and lichens may be an unfavorable result. Mineral accretions, including ferromanganese crusts, may be deposited from solute-rich solutions. Rainwater is never pure water; it acquires numerous ions already before it reaches the ground, many of which assist it in converting or dissolving minerals. There is a distinct tendency for such inclusions, which are usually chlorides, sulfates, nitrates, phosphates and bicarbonates, to effect a lowering of the atmospheric water pH. It is well known that this has been exacerbated by the industrial development of recent centuries ('acid rain'). However, organic acids are also present in rainwater. After noting the high contents of formate, acetate and proprionate in cloud and rainwater in tropical Australia (Noller et al. 1987), which is attributable to large-scale vegetation burning, Watchman (1990) suggested that these acids might cause the formation of oxalate crusts. Isoprene, found in many plants, is released into the atmosphere, converted by hydroxyl radicals to formaldehyde, which is then oxidized to formic acid. Similar photochemical reactions can convert other breakdown products of isoprene into oxalic acid, which can convert carbonates, including some rock art pigments, to oxalates.

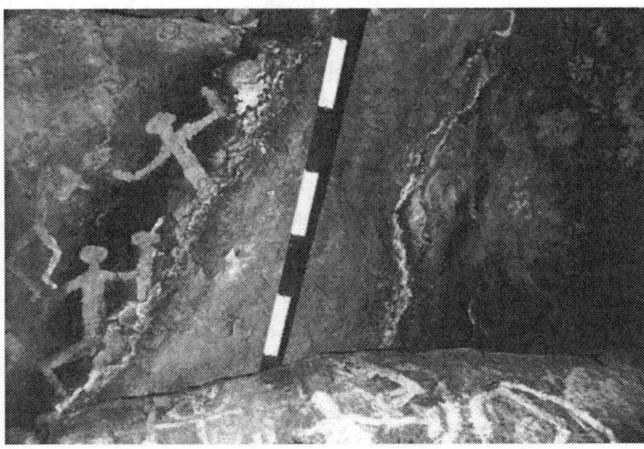

Figure 14. Pictograms affected by run-off water, Mt Grenfell, central NSW, Australia.

Mass-exfoliation of discrete rock facies in stratified deposits can be caused by either gravitational or capillary moisture traveling within the rock and selectively degrading specific strata. It is one of the more pernicious rock art conservation problems, because it not only destabilizes otherwise perhaps quite well-preserved panels, but it is particularly difficult to address. Attempts to control capillary water, in particular, have a long history of failure. A comparatively stable rock art panel may be disintegrating because it depends on the structural support of another, less stable facies that is rapidly dissolving through moisture. Two different regimes are known, one exemplified by Shishkino in central Siberia, where the rock art occurs on a vertical cliff, on a horizontal band of sound sandstone. Immediately below this stratum is a layer of argillaceous sandstone (almost mudstone), which decays rapidly from the action of internal water seepage. As it disintegrates, it leaves the stratum without support, which will eventually collapse from its own weight (Bednarik and Devlet 1993). In the second case, at Mootwingee in eastern Australia, a steep slope dipping at about 19° consists of relatively resistant siliceous sandstone. It bears the rock art, but is underlain by a layer cemented mainly by gypsum and hydrated aluminum silicates, both of which are removed readily by chemical weathering. The upper layer develops fissures as a result, which in turn increases access by water and weathering of the lower stratum. Because of the slope's steepness individual blocks, which 'float' on this increasingly unstable support, then slide downwards. While this is seen as a form of physical weathering, the true source of the deterioration is primarily chemical action of moisture. Similar effects can work at a smaller scale, for instance Ollier's (1969) 'dirt cracking', which is essentially attributable to the swelling of clays deposited in fine rock fissures.

Physical Weathering

While moisture does cause a variety of physical rock deterioration processes, such as flaking or exfoliation, other such processes involve no chemical changes wrought by moisture. They can have devastating effects on rock art and are often much easier to alleviate. While Rosenfeld (1985) lists hydration, 'salt decay' and the removal of cementing medium, here the term physical rock weathering is limited to purely physical processes.

Insolation or solar radiation is capable of breaking up some rock types through thermal expansion and contraction, particularly in arid or semi-arid regions experiencing significant diurnal temperature variations (Goodwin 1960; Ollier 1969). Once controversial, the effectiveness of this process has been shown through the buckling of marble slabs (Kessler 1919). The ubiquitous sub-spherical rocks of some deserts are largely the result of temperature stresses over great time spans. The rounding, most especially of granite, is the result of the differential temperature transference within the rock, which heats or cools protruding aspects or corners more readily than planar surfaces (see above). Over time the repetitiveness of the process along a stress zone leads to eventual material fatigue. Ideally the process would lead to a spherical shape (or more precisely, a body with the smallest possible surface area relative to mass), and the spalls are always of distinctly convex fracture surfaces. The process is assisted by fire damage, which has the same effect but yields quite different spalls. Thermal stress fatigue is believed to be responsible for a variety of fracturing patterns other then the described spherical reduction, which have also affected rock art sites, particularly in combination with other factors (Hall and Hall 1991; Meiklejohn 1997). The extensive exfoliation of the cupule-bearing walls of Daraki-Chattan Cave in India, one of the world's oldest known rock art sites, is conspicuously limited to the panels accessible to the sun, and is almost certainly attributable to insolation (Bednarik et al. 2005).

Lightning strikes can result in fracturing of boulders of any size, and such traces are easy to recognize (Figure 15). This is a major cause of rock reduction, especially in flat areas where rocky outcrops tend to be the only elevated features. Boulders at or near the peak or along upper ridges are favored targets of lightning, which usually strikes the uppermost surface of boulders. A rock may be broken in half, or have large fragments flaked off. At the point of impact the rock may be shattered and its surface is commonly glazed where surface material has became fused. If there are iron-rich rinds present, the minerals formed under very high temperatures of the electrical discharges tend to be of a light greenish to bluish color (Bednarik 1979).

Figure 15. Effect of recent lightning strike on petroglyph-covered granite boulder, Long Tin Creek site, Pilbara, Western Australia.

Kernsprung is a phenomenon of boulder fracture whose results resemble those of lightning strikes, but there is no impact apparent, nor any of the other features just described. The cause is not clearly established, but reports of extremely loud, explosion-like sounds in arid areas suggest that cumulative temperature-induced stresses could build up to the point of a spontaneous release, causing an apparently structurally sound boulder to split literally in half.

Brushfires are a frequent agent of physical weathering (Blackwelder 1927; Emery 1944; Gunn and Whear 2009; Gunn 2011), especially on rock types that are less susceptible to other forms of weathering (Figure 1). The heat developed by the burning of resinous vegetation in arid regions can be quite intense (temperatures of 800°C and much higher have been reported; Rosenfeld 1985: 33; Gunn 2011), resulting in rapid heating of a thin surface zone, and fatigue between this and the core mass of the boulder which is unaffected by this sudden temperature change. Due to the geometry of heat transference in a solid body, which obeys the same fundamental laws as wane formation as enunciated above, protruding aspects heat up more rapidly and the flakes detached are typically convex and under 20 mm maximum thickness (Selkirk and Adamson 1981), although thicker specimens have been observed.

Another process causing shattering of rock is attributable to ice formation, or to freeze-and-thaw cycles. Liquid water is virtually incompressible, having its greatest density at +4°C; frozen, it expands by 1/11th of its volume. However, crystallization of water as such does not cause fracturing of rocks, it can only exacerbate already existing cracking by penetrating fissures. In cracks that are less than a few tenth of a millimeter wide, even temperatures as low as -20°C will not induce water to relinquish its liquid state (Bednarik 1979: 16). On the other hand if openings exceed a few millimeters, congelifraction is not effective because the expansion of water through crystallization will merely force water out. Rock fatigue due to freeze-and-thaw cycles is perhaps attributable to either the compression of air sealed off in voids by the developing ice (Tricart 1969) or to regelation (Schmid 1958). The extremely high content of cryoclasts in stadial, entrance-near deposits in caves shows tellingly how effective congelifraction is on at least some rock types. Porous rocks (limestone and dolomite, for instance) are subjected to severe fracturing (Bednarik 1970), as are schists because of their laminate structure inviting penetration by gravitational water.

Tectonic changes in rock facies or rock structures can also affect rock art sites in a number of ways. These are structural responses, usually due to gravitational disequilibrium of some form. Not all such changes are attributable to natural causes, for instance subsidence due to underground extraction of coal is obviously an anthropic threat. Natural tectonic causes that have destroyed rock art include seismic activity, relocation of major sediment deposits which render rock structures unstable, and the effects of aquifer level fluctuations (in partially water-filled cave systems). Petroglyphs often occur near rivers or at coasts, on bedrock or boulders which are exposed to yet another form of physical damage, due to the kinetic energy of water. They may be moved by the water, or sand moved by the water may act as an abrasive on them (Bednarik 2009a, 2010). Sandblasting is not, however, limited to water-propelled quartz sand, it also effects aeolian erosion. While this is barely effective on hard and structurally sound rock, *aeolian erosion* is highly effective on softer rock types, most especially on calcarenite, other soft sandstones, even limestone, especially in coastal and arid environments of high wind velocity. Sandblasting can destroy pictograms and even deep petroglyphs in a matter of a few decades. These weathering processes are often facilitated by weakening of the rock fabric through moisture or salt action.

Biological Weathering Factors

Non-human biological agents involved in the deterioration of rock art panels range from large mammals through to microfauna, from large plants through to cryptogamic growths, and include bacteria. All rock surfaces bear biotic substances, and all rock substrates contain organic carbon (Bednarik 1979), which as we have seen above is the principal reason why carbon dating of rock surface features only yields questionable results in many circumstances. Rock surfaces are sometimes implicitly regarded as somehow 'sterile', when in fact bacteria occur even several kilometers deep in the lithosphere. Microorganisms are also implicated in the formation and re-cycling of specific accretionary deposits as well as in the removal of various cations from numerous mineral components. Not only do such activities contribute to the weakening of rock fabric (Berthelin 1988), they are contingent upon the production of corrosive agents. Their metabolic products may also be corrosive (Eckhardt 1978). Broadly speaking, we are concerned with bacteria utilizing atmospheric nitrogen (converting it to ammonia, or oxidizing this to nitrous or nitric acid); those oxidizing sulfur to sulfur dioxide and converting it to highly corrosive sulfuric acid or the weaker sulfurous acid; those oxidizing iron and manganese from minerals; and those attacking silicates and phosphates with organic acids, in particular ketone-based compounds. However, two points need to be considered here: first, the quantities of such highly effective acids are extremely small; second, the processes they cause are not necessarily only corrosive. For instance *Ferrobacillus* sp. or *Thiobacillus* sp. do consume cations, but often from airborne material rather than the host rock, and they may accrete protective surface crusts that actually preserve petroglyphs. Similarly, the mobilization of amorphous silica, for instance through 2-ketogluconic acid produced by bacteria, can lead to the deposition of silica skins, which have been observed to assist in consolidating both petroglyphs and pictograms. Therefore microbial action is not readily identifiable in most cases, nor are the processes readily definable.

The second group of Thallophyta, the fungi, also require no chlorophyll and can exist in complete darkness. By producing oxalic or citric acid, some species can attack various minerals more aggressively than bacteria, except calcium-rich minerals which seem to be more susceptible to bacterial attack. The citric acid of fungi lead to hydrolysis of silicate minerals (Silverman and Muñoz 1970; Winkler 1973). In contrast to fungi, algae require light for photosynthesis, although filtered and weak light suffices for some species. While algae range from single-celled forms to very complex multicellular species, two types are of particular interest to the rock art conservator. The blue-green algae (Cyanophyta), because of their involvement in the production of rock varnish (Scheffer et al. 1963), and the terrestrial green algae (Chlorophyta), which like the former class occur in lichen. Algae do not seem to attack rock surfaces directly, but they can have a damaging effect through their ability of moisture retention, and they depend on the presence of moisture.

Lichens result from a symbiosis between fungi, usually of the class Ascomycetes, and blue-green or green algae. There is very limited evidence of real damage lichens have caused to petroglyphs. Some authors even suggest that a lichen cover may reduce rock weathering rates (Wilson 1995), or while acknowledging deleterious effects have warned that removal of lichen may exacerbate substrate deterioration (Florian 1978). Jackson and Keller's (1970) study, while showing that the weathering zone under lichen cover is thicker than on nearby rock that is free of lichen, is inconclusive, because it cannot tell us whether this is because the lichen has prevented erosion, or because the weathering zone develops faster under the lichen. Cochran and Berner (1992, 1993) report the preservation of a silica skin beneath colonizing lichens, while other authors suggest that lichens could enhance rates of weathering (Fry 1924, 1927; Syers and Iskandar 1973).

Mosses are more effective in retaining moisture for extended periods of time and are thus probably a greater conservation threat. They are non-flowering plants lacking roots, but whose stems are equipped with rhizoids to anchor them. Because of their often much greater bulk in comparison to lichens, mosses attract airborne debris and edaphic sediment to assist in colonizing rock surfaces, which conflicts with the survival of rock art. Even larger plants, such as Gramineae, shrubs and trees can become major agents of rock weathering. Large tufts of grass or plant branches rubbing against relatively soft rock surfaces over years produce kinetic rock marks, as do tree roots hugging rocks for support and moving slightly as the tree sways in the wind (Bednarik 1994b).

Insect species also contribute to the biological weathering at rock art panels. Termites (Isoptera) cause damage to rock paintings through their construction of 'runways' (Watson and Flood 1987). In their construction, sand grains are removed from a supporting sandstone panel and placed on the surface of the runway sheath. The nests of numerous species of mud-daubing wasps (Hymenoptera) have been observed at rock art sites, as various different structures (e.g. tubular, sub-hemispherical, polygonal). Where bonding of the nest structure by mineral precipitation has occurred, exfoliation removes some of the host rock's surface. But there is no evidence that these insects erode the rock surface, therefore the damage they inflict on rock art is limited to the exfoliation phase of their nests (Naumann and Watson 1987). Some species of bees burrow into soft rock types, such as calcarenite or weathered sandstone. Such nesting aggregations can include hundreds of closely adjacent entrances, each several millimeters in diameter.

Birds damage rock art through the acidity of their droppings, which facilitates dissolving of rock varnish and other protective deposits (Bednarik 1979) and establishment of cryptogamic growths. In regions of deeply patinated rock outcrops, the highest points are inevitably free of patination, because they are the favorite perches of birds. Therefore discoloration of such an aspect is not necessarily evidence of lightning strike. Numerous larger animals contribute to biological weathering of rock art, especially through rubbing their bodies against shelter walls, but also by adding their claw markings, especially in caves. This includes the traces of now extinct megafauna, such as mammoth rubbings in North America and polished walls in the caves of Europe where countless generations of cave bears smoothed walls with their bodies, with sand embedded in their fur acting as an abrasive (Bednarik 1993).

One single mammalian species, however, has managed to eclipse the combined effects on rock art of all other forms of biological weathering. Quite probably, *Homo sapiens sapiens* has even outperformed chemical and physical weathering in the destruction of rock art, especially through development projects, and to a lesser degree through iconoclasm, tourism and misguided research practices. Although the destructive activities of man amount to biological weathering when they affect rocks, this 'anthropic weathering' is of such magnitude that it cannot be explored adequately in the present context. Suffice it to say that the consequences sustained by rock art through the one species whose footprint on this planet rivals that of the impact of a major meteorite have been decidedly massive.

Robert G. Bednarik
International Federation of Rock Art Organizations (IFRAO)
P.O. Box 216
Caulfield South, VIC 3162
Australia
robertbednarik@hotmail.com

References

Acker, J. G., & Bricker, O. P. (1992) The influence of pH on biotite dissolution and alteration kinetics at low temperature. *Geochimica et Cosmochimica Acta* 56: 3073–3092.

Ampoulange, U., & Pintaud, R. C. (1955) Une nouvelle gravure de la grotte de La Grèze (Dordogne). *Bulletin de la Société Préhistorique Française* 52: 249–251.

Anati, E. (1976) *Evolution and style in Camunian rock art*. Capo di Ponte: Archivi 6.

Aoki, F. (1951) Molecular dissolution of silica in water. *Journal of the Chemistry Society of Japan, Pure Chemistry Section* 72: 17.

Armitage, R. A., Brady, J. E., Cobb, A., Southon, J. R., & Rowe, M. W. (1999) Radiocarbon dates on charcoal pigments from three Mayan glyph texts in Naj Tunich Cave, Guatemala. Unpubl. MS.

Atkinson, T. C., & Smith, D. I. (1976) The erosion of limestones. In T. D. Ford & C. H. D. Cullingford (Eds), *The science of speleology*, pp. 151–177. London: Academic Press.

Bazylinski, D. A., & Frankel, R. B. (2003) Biologically induced mineralization by bacteria. *Reviews in Mineralogy and Geochemistry* 54: 95–114.

Bednarik, R. G. (1970) Die Grabungen in der Promenadensteighöhle (1961–1964). *Die Höhle* 21: 117–126.

Bednarik, R. G. (1979) The potential of rock patination analysis in Australian archaeology — part 1. *The Artefact* 4: 14–38

Bednarik, R. G. (1980) The potential of rock patination analysis in Australian archaeology — part 2. *The Artefact* 5: 47–77.

Bednarik, R. G. (1981) Finger lines, their medium, and their dating. Unpubl. MS, 34 pp., Archive of the Australian Rock Art Research Association, Melbourne.

Bednarik, R. G. (1984) Die Bedeutung der paläolithischen Fingerlinientradition. *Anthropologie* 23: 73–79.

Bednarik, R. G. (1987) No pictographs at end of Rochester Creek rainbow. *La Pintura* 15(2+3): 14–18.

Bednarik, R. G. (1989a) Perspectives of Koongine Cave and scientistic archaeology. *Australian Archaeology* 29: 9–16.

Bednarik, R. G. (1989b) On the Pleistocene settlement of South America. *Antiquity* 63: 101–111.

Bednarik, R. G. (1992) A new method to date petroglyphs. *Archaeometry* 34: 279–291.

Bednarik, R. G. (1993a) A strategy for the practical application of microerosion dating. In J. Steinbring & A. Watchman (Eds), *Time and space: Proceedings of Symposium F, Second AURA Congress*, pp. 64–66. Melbourne: Occasional AURA Publication 8, Australian Rock Art Research Association.

Bednarik, R. G. (1993b) Geoarchaeological dating of petroglyphs at Lake Onega, Russia. *Geoarchaeology* 8: 443–463.

Bednarik, R. G. (1994a) Conceptual pitfalls in dating of Palaeolithic rock art. *Préhistoire Anthropologie Méditerranéennes* 3: 95–102.

Bednarik, R. G. (1994b) The discrimination of rock markings. *Rock Art Research* 11(1): 23–44.

Bednarik, R. G. (1995) The age of the Côa valley petroglyphs in Portugal. *Rock Art Research* 12: 86–103.

Bednarik, R. G. (1996) Only time will tell: a review of the methodology of direct rock art dating. *Archaeometry* 38: 1–13.

Bednarik, R. G. (1997) Microerosion analysis of petroglyphs in Valtellina, Italy. *Origini* 21: 7–22.

Bednarik, R. G. (1998) Cosmogenic radiation nuclides in archaeology: a response to Phillips et al. *Antiquity* 72: 811–815.

Bednarik, R. G. (1999) The speleothem medium of finger flutings and its isotopic geochemistry. *The Artefact* 22: 49–64.

Bednarik, R. G. (2000) Some problems with 'direct dating' of rock-pictures. In G. K. Ward & C. Tuniz (Eds), *Advances in dating Australian rock-*

markings, pp. 104–109. Melbourne: Occasional AURA Publication 10, Australian Rock Art Research Association.

Bednarik, R. G. (2000b) Age estimates for the petroglyph sequence of Inca Huasi, Mizque, Bolivia. *Andean Past* 6: 277–287.

Bednarik, R. G. (2001a) The taphonomy of beeswax figures. *Rock Art Research* 18: 91–95.

Bednarik, R. G. (2001b) *Rock art science: the scientific study of palaeoart*. Turnhout: Brepols (2nd edition 2007, New Delhi: Aryan Books International).

Bednarik, R. G. (2002a) About the age of Pilbara rock art. *Anthropos* 97(1): 201–215.

Bednarik, R. G. (2002b) The dating of rock art: a critique. *Journal of Archaeological Science* 29(11): 1213–1233.

Bednarik, R. G. (2007) The science of Dampier rock art — part 1. *Rock Art Research* 24(2): 209–246.

Bednarik, R. G. (2009a) Fluvial erosion of inscriptions and petroglyphs at Siega Verde, Spain. *Journal of Archaeological Science* 36(10): 2365–2373.

Bednarik, R. G. (2009b) Experimental colorimetric analysis of petroglyphs. *Rock Art Research* 26(1): 55–64.

Bednarik, R. G. (2010) Australian rock art of the Pleistocene. *Rock Art Research* 27(1): 95–120.

Bednarik, R. G. (2011) *The human condition*. New York: Springer.

Bednarik, R. G. (in prep.) The age estimation of mineralised wasp nests.

Bednarik, R. G., Bentley, J., & Carpenter, D. (in press) The Zuitdorp inscriptions, Western Australia. *Rock Art Research* 28.

Bednarik, R. G., & Devlet, E. G. (1993) Problemy konservatsiï pamyatnikov naskal'nogo iskusstva verkhneï Leny. *Sovremennye problemy isucheniya petroglifov*, pp. 37–48. Kemerovo: Spornik Nauchnykh Trudov, Kemerovskiï Rocudarstvennyï Universitet.

Bednarik, R. G., Kumar, G., Watchman, A., & Roberts, R. G. (2005) Preliminary results of the EIP Project: the status at the end of 2004. *Rock Art Research* 22(2): 147–197.

Belzoni, G. (1820) *Narrative of the operations and recent discoveries within the pyramids, temples, tombs and excavations in Egypt and Nubia*, Vol. 1. London.

Berthelin, J. (1988) Microbial weathering processes in natural environments. In A. Lerman & M. Meybeck (Eds), *Physical and chemical weathering in geochemical cycles*, pp. 33–59. New York: Kluwer Academic

Publishers.

Birkeland, P. W., Colman, S. M., Burke, R. M., & Mierding, T. C. (1979) Nomenclature of alpine glacial deposits, or, what's in a name? *Geology* 7: 532–536.

Blackwelder, E. (1927) Fire as an agent of rock weathering. *Journal of Geology* 35: 134–140.

Bosazza, V. L. (1937) A study of the chert of the dolomitic series of the Transvaal system. *South African Journal of Science* 34: 178–185.

Brandl, E. (1968) Aboriginal rock designs in beeswax and description of cave painting sites in western Arnhem Land. *Archaeology and Physical Anthropology in Oceania* 3: 19–29.

Bretz, J. H. (1942) Vadose and phreatic features of limestone caves. *Journal of Geology* 50(6): 675–811.

Burke, R. M., & Birkeland, P. W. (1979) Re-evaluation of multiparameter relative dating techniques and their application to the glacial sequence along the eastern escarpment of the Sierra Nevada, California. *Quaternary Research* 11: 21–51.

Busenberg, E., & Clemency, C. V. (1976) The dissolution kinetics of feldspars at 25°C and at 1 atm CO_2 partial pressure. *Geochimica et Cosmochimica Acta* 40: 41–50.

Cailleux, A. (1951) Morphoskopische Analyse der Geschiebe und Sandkörner und ihre Bedeutung für die Paläoklimatologie. *Geologische Rundschau* 40: 5–13.

Campbell, I. A. (1991) Classification of rock weathering at Writing-on-Stone Provincial Park, Alberta, Canada: a study in applied geomorphology. *Earth Surface Processes and Landforms* 16: 701–711.

Campbell, J. B. (2000) The Chillagoe and Laura laser-AMS dating project. In G. K. Ward & C. Tuniz (Eds), *Advances in dating Australian rock-markings*, pp. 80–83. Melbourne: Occasional AURA Publication 10, Australian Rock Art Research Association.

Cannon, W. J., & Ricks, M. J. (1986) The Lake County, Oregon, rock art inventory: implications for prehistoric settlement and land use patterns. *Contributions to the Archaeology of Oregon* 3: 1–23.

Capitan, L., Breuil, H., & Peyrony, D. (1912) Les gravures sur cascade stalagmitique de la grotte de la Mairie à Teyat (Dordogne). *Congrès International d'Anthropologie et d'Archéologie Préhistorique, 16e session*, pp. 498–514. Geneva.

Carroll, T. (1974) Relative age dating techniques and a Late Quaternary chronology, Arikaree Cirque, Colorado. *Geology* 2: 321–325.

Černohouz, J., & Solč, I. (1966) Use of sandstone wanes and weathered basaltic crust in absolute chronology. *Nature* 212: 806–807.

Chaloupka, G. (1993) *Journey in time*. Reed Books, Chatswood.

Chou, L., Garrels, R. M., & Wollast, R. (1989) Comparative study of the kinetics and mechanisms of dissolution of carbonate minerals. *Chemistry and Geology* 78: 269–282.

Clark, D. J. (1979) The Gambieran stone tool industry. Unpubl. B.A. thesis, Division of Prehistory, La Trobe University.

Cochran, M. F., & R. A. Berner (1992) The quantitative role of plants in weathering. In Y. K. Kharaka & A. S. Maest (Eds), *Water-rock interaction*, pp. 473–476. Rotterdam: Balkema.

Cochran, M. F., & R. A. Berner (1993) Reply to the comments on 'Weathering, plants, and the long-term carbon cycle'. *Geochimica et Cosmochimica Acta* 57: 2147–2148.

Cole, D. R., & Monger, H. C. (1994) Influence of atmospheric CO_2 on the decline of C4 plants during the last deglaciation. *Nature* 368: 533–36.

Colman, S. M. (1981) Rock-weathering rates as functions of time. *Quaternary Research* 15: 250–264.

Colman, S. M., & Pierce, K. L. (1981) *Weathering rinds on andesitic and basaltic stones as a Quaternary age indicator, western United States*. U.S. Geological Survey, Professional Paper 1210.

Consens, M. (2006) Between artefacts and egofacts: the power of assigning names. *Rock Art Research* 23(1): 79–84.

Correns, C. W., & Nagelschmidt, G. (1933) Über Faserbau und optische Eigenschaften von Chalzedon. *Zeitschrift für Kristallographie* 85: 199–213.

Crivelli Montero, E. A., & Fernández, M. M. (1996) Palaeoindian bedrock petroglyphs at Epullán Grande Cave, northern Patagonia, Argentina. *Rock Art Research* 13: 112–117.

Crook, R. (1986) Relative dating of Quaternary deposits based on P-wave velocities in weathered granitic clasts. *Quaternary Research* 25: 281–292.

Curl, R. L. (1966) Scallops and flutes. *Transactions of the Cave Research Group of Great Britain* 7(2): 121–160.

Curl, R. L. (1974) Deducing flow velocity in cave conduits from scallops. *Bulletin of the National Speleological Society* 36(2): 1–5.

Curwen, E. C. (1940) The white patination of black flint. *Antiquity* 14: 435–437.

Daleau, F. (1896) Les gravures sur roche de la caverne de Pair-non-Pair.

Actes de la Société Archeologique de Bordeaux 21: 235.

David, B., Armitage, R. A., Hyman, M., Rowe, M. W., & Lawson, E. (1999) How old is north Queensland rock art? A review of the evidence, with new AMS determinations. *Archaeology in Oceania* 34: 103–120.

David, P. (1934) Abri de la Chaire à Calvin. *Congrès Préhistoire de France, 11e session,* pp. 372–378. Périgueux.

Davis, R., & Schaeffer, O. A. (1955) Chlorine-36 in nature. *Annals of the New York Academy of Sciences* 62: 107–121.

Day, M. J., & Goudie, A. S. (1977) Field assessment of rock hardness using the Schmidt Hammer. *British Geomorphological Research Group Technical Bulletin* 18: 19–29.

Del Monte, M., & Sabbioni, S. (1987) A study of the patina called 'scialbatura' on Imperial Roman marbles. *Studies in Conservation* 32: 114–121.

De Serres, P. M. T. (1835) *Essai sur les caverns à ossemens et sur les causes qui les y ont accumulés*. Haarlem: Loosjes.

Denninger, E. (1971) The use of paper chromatography to determine the age of albuminous binders and its application to rock paintings. Rock paintings in southern Africa, supplement to the *South African Journal of Science*, Special Issue, 2: 80–84.

Diénert, F., & Wandenbulcke, F. (1923) Sur le dosage de la silice dans les eaux. *Comptes Rendus* 176: 1487–1480.

Diénert, F., & Wandenbulcke, F. (1924) Une étude de la colloid silice. *Comptes Rendus* 178: 564–566.

Donaldson, M. (2011) Understanding the rocks: rock art and the geology of Murujuga (Burrup Peninsula). *Rock Art Research* 28(1).

Donnay, J. D. H. (1936) La birefringence de forme dans la chalcedonie. *Société Geologique de Belgique Annales* 59: 289–302.

Dorn, R. I. (1983) Cation-ratio dating: a new rock varnish age-determination technique. *Quaternary Research* 20: 49–73.

Dorn, R. I. (1986) Rock varnish as an indicator of aeolian environmental change. In W. G. Nickling (Ed.), *Aeolian geomorphology*, pp. 291–307. London: Allen & Unwin.

Dunnell, R. C., & Readhead, M. L. (1988) The relation of dating and chronology: comments on Chatters and Hoover (1986) and Butler and Stein (1988). *Quaternary Research* 30: 232–233.

Emery, K. G. (1944) Brush fires and rock exfoliation. *American Journal of Science* 242: 506–508.

Emery, K. G. (1960) Weathering of the Great Pyramid. *Journal of*

Sedimentary Petrology 30: 140–143.

Engel, C. E., & Sharp, R. P. (1958) Chemical data on desert varnish. *Bulletin of the Geological Society of America* 69: 487–518.

Eckhardt, F. E. W. (1978) Microorganisms and weathering of a sandstone monument. In W. E. K. Krumbein (Ed.), *Environmental biochemistry and geomicrobiology*, pp. 675–686. Ann Arbor: Ann Arbor Science Publishers.

Fairbridge, R. W. (1964) The importance of limestone and its Ca/Mg content to palaeoclimatology. In A. E. M. Nairn (Ed.), *Problems in palaeoclimatology*, pp. 431–478. London.

Fetterman, J. (1996) Radiocarbon and tree-ring dating at early Navajo sites. In R. H. Towner (Ed.), *The archaeology of Navajo origins,* pp. 71–82. Salt Lake City: University of Utah Press.

Florian, M. L. E. (1978) A review: the lichen role in rock art — dating, deterioration and control. In C. Pearson (Ed.), *Conservation of rock art*, pp. 95–98. Sydney: Institute for the Conservation of Cultural Material.

Folk, R. L., & Weaver, C. E. (1952) A study of the texture and composition of chert. *American Journal of Science* 250(7): 498–510.

Francis, J., Loendorf, L. L., & Dorn, R. I. (1993) AMS radiocarbon and cation-ratio dating of rock art in the Bighorn Basin of Wyoming and Montana. *American Antiquity* 58: 711–737.

Fry, E. J. (1924) A suggested explanation of the mechanical action of lithophytic lichens on rocks (shale). *Annals of Botany* 38: 175–196.

Fry, E. J. (1927) The mechanical action of crustaceous lichens on substrata of shale, schist, gneiss, limestone, and obsidian. *Annals of Botany* 41: 437–460.

Fullagar, R. L. K., Price, D. M., & Head, L. M. (1996) Early human occupation of northern Australia: archaeology and thermoluminescence dating of Jinmium rock-shelter, Northern Territory. *Antiquity* 70: 751–773.

Gehrke, E. (1935) Die Patina von norddeutschem Flint. *Zeitschrift für Geschiebeforschung* 2: 107–114.

Gellatly, A. F. (1984) The use of rock weathering-rind thickness to redate moraines in Mount Cook National Park, New Zealand. *Arctic and Alpine Research* 16: 225–232.

Goodwin, A. J. H. (1960) Chemical alteration (patination) of stone. In R. F. Heizer & S. F. Cook (Eds), *The application of quantitative methods in archaeology*, pp. 300–312. New York: Viking Fund Publications in Anthropology 28.

Gunn, R. G. (2011) The impact of bushfires and fuel reduction burning on the preservation of shelter rock art. *Rock Art Research* 28(1): 53–69.

Gunn, R. G., & Whear, R. L. (2009) Bushfire-induced heat and smoke patterns within an Aboriginal rock art shelter. *Rock Art Research* 26(2): 205–212.

Hale, H. M., & Tindale, N. B. (1930) Notes on some human remains in the lower Murray valley, South Australia. *Records of the South Australian Museum* 4: 145–218.

Harman, R. W. (1926) Aqueous solutions of sodium silicates. *Journal of Physical Chemistry* 30: 1100–1111.

Hedges, R. E. M., Ramsey, C. B., Van Klinken, G. J., Pettitt, P. B., Nielsen-Marsh, C., Etchegoyen, A., Fernandez Niello, J. O., Boschin, M. T., & Llamazares, A. M. (1998) Methodological issues in the ^{14}C dating of rock paintings. *Radiocarbon* 40: 35–44.

High, C., & Hanna, F. K. (1970) A method for the direct measurement of erosion of rock surfaces. *British Geomorphological Research Corps, Technical Bulletin* 5: 24.

Hall, K. J., & Hall, A. (1991) Thermal gradients and rock weathering at low temperatures: some simulation data. *Permafrost and Periglacial Processes* 2: 103–112.

Hurst, V. J., & Kelly, A. R. (1966) Patination of cultural flints. In J. R. Caldwell (Ed.), *New roads to yesterday*, pp. 517–523. London.

Iler, R. K. (1955) *Colloid chemistry of silica and silicates*. Ithaca: Cornell University Press.

Jackson, T. A., & Keller, W. D. (1970) A comparative study of the role of lichens and 'inorganic' processes in the chemical weathering of recent Hawaiian lava flows. *American Journal of Science* 269: 446–66.

Jolles, H., & Neurath, F. (1898) Colorimetric estimation of silica in water. *Zeitschrift für Angewandte Chemie* 11: 315–316.

Judd, J. W. (1887) On the making of flints. *Proceedings of the Geologists Association* 10: 217–226.

Kahlenberg, L., & Lincoln, A. T. (1898) Solutions of silicates of the alkalies. *Journal of Physical Chemistry* 2: 77–90.

Kappler A., Pasquero, C., Konhauser, K. O., & Newman, D. K. (2005) Deposition of banded iron formations by anoxygenic phototrophic Fe(II)-oxidizing bacteria. *Geology* 3: 865–868.

Keller, W. D. (1957) *The principles of chemical weathering*. Columbia, Miss.

Kessler, D. W. (1919) *Physical and chemical tests on commercial marbles of the United States*. Technological Papers of the U.S. Bureau of Standards

No. 123.

Konhauser, K. O. (1997) Bacterial iron biomineralisation in nature. *FEMS Microbiology Reviews* 20*:* 315–326.

Krauskopf, K. B. (1956) Dissolution and precipitation of silica at low temperatures. *Geochimica et Cosmochimica Acta* 10: 1–26.

Lalanne, G., & Breuil, H. (1911) L'abri sculpté du Cap-Blanc à Laussel, Dordogne. *L'Anthropologie* 21: 385–402.

Layton, R. (1991) Figure, motif and symbol in the hunter-gatherer rock art of Europe and Australia. In P. Bahn & R. Rosenfeld (Eds), *Rock art and prehistory*, pp. 23–38. Oxford: Monograph 10, Oxbow Books.

Layton, R. (2001) Ethnographic study and symbolic analysis. In D. S. Whitley (Ed.), *Handbook of rock art research*, pp. 311–331. Altamira Press, Walnut Creek, California.

Lazzarini, L., & Salvadori, O. (1989) A reassessment of the formation of the patina called *scialbatura*. *Studies in Conservation* 34: 20–26.

Lemozi, A. (1920) Peintures et gravures paléolithiques découvertes dans les grottes des communes d'Espagnac, de Sainte-Eulalie et de Cabrerets. *Bulletin de la Société Préhistorique Française* 17: 256–263.

Lin, F., & Clemency, C. V. (1981) The kinetics of dissolution of muscovites at 25°C and 1 atm CO_2 partial pressure. *Geochimica et Cosmochimica Acta* 45: 571–576.

Lüttig, G. (1956) Eine neue, einfache geröllmorphometrische Methode. *Eiszeitalter und Gegenwart* 7: 13–20.

Mabbutt, J. A. (1961) 'Basal surface' or 'weathering front'. *Proceedings of the Geological Association in London* 72: 357–358.

McCarroll, D. (1991) The Schmidt Hammer, weathering and rock surface roughness. *Earth Surface Processes and Landforms* 16: 477–480.

McCarthy, D., Payen, L., & Ennis, P. (1994) Use of amino acid racemization methods on rock paintings at Motte Rimrock Reserve. *American Indian Rock Art* 13: 17–22.

McDonald, J., Officer, K., Jull, T., Donahue, D., Head, J., & Ford, B. (1990) Investigating ^{14}C AMS: dating prehistoric rock art in the Sydney Sandstone Basin, Australia. *Rock Art Research* 7: 83–92.

Meiklejohn, K. I. (1997) Rock art sites, deterioration, protection and utilisation. Paper presented to conference on 'Learning and teaching about rock art and the people who made it', African Window, Pretoria, 19 April 1997.

Midgley, H. G. (1951) Chalcedony and flint. *Geological Magazine* 88: 179–184.

Mihevc, A., Slabe, T., & Sebel, S. (2004) The morphology of caves. In J. Gunn (Ed.), *Encyclopedia of caves and karst science*, pp. 521–124. New York: Fitzroy Dearborn.

Milner, H. B. (1962) *Sedimentary petrography*, 4th edition. London.

Mitchell, S. R. (1947) Patination. *Mankind* 3: 298–302.

Moore, E. S., & Maynard, J. E. (1929) Solution, transportation and precipitation of iron and silica. *Economic Geology* 24: 272–303.

Mosebach, R. (1955) Die hydrothermale Löslichkeit des Quarzes als heterogenes Gasgleichgewicht. *Neues Jahrbuch mineralogischer Abhandlungen* 87: 351–388.

Mulvaney, J. D. (1969) *The prehistory of Australia*. London: Thames & Hudson.

Mulvaney, K. (1995) What to do on a rainy day. Reminiscences of Mirriuwung and Gadjerong artists. *Rock Art Research* 13: 3–20.

Murray, J., & R. Irvine (1891) On silica and siliceous remains of organisms in modern seas. *Proceedings of the Royal Society of Edinburgh* 18: 229–250.

Muzzolini, A. (1999) La notation des dates au 14 C. *Cahiers de l'AARS* 16: 18–24.

Naumann, I. D., & Watson, J. A. L. (1987) Wasps and bees (Hymenoptera) on rock faces at Koolburra. *Rock Art Research* 4: 26–27.

Nelson, D. E. (1993) Second thoughts on a rock-art date. *Antiquity* 67: 893–895.

Nelson, D. E. (Ed.) (2000) *The beeswax art of northern Australia*. Compact disk, Simon Fraser University, Burnaby, Canada.

Nelson, D. E., Chaloupka, G., Chippindale, C., Alderson, M. S., & Southon, J. R. (1995) Radiocarbon dates for beeswax figures in the prehistoric rock art of northern Australia. *Archaeometry* 37: 151–156.

Nobbs, M. F., & Dorn, R. I. (1988) Age determinations for rock varnish formation within petroglyphs: cation-ratio dating of 24 motifs from the Olary region, South Australia. *Rock Art Research* 5: 108–146.

Noller, B. N., N. Currey, G. Ayers, R. Gillett, A. Harrison & R. A. Rasmussen (1987) Chemistry of rain-water acidity, nutrients and trace metals. In *Alligator Rivers Region Research Institute Annual Report Summary for 1985–86*, pp. 53–55. Canberra: Australian Government Publishing Service.

Novellino, D. (1999) Towards an understanding of Pälaqwan rock drawings: between visual and verbal expression. *Rock Art Research* 16: 3–24.

Oakley, K. P. (1939) The nature and origin of flint. *Science Progress* 34:

277–286.

O'Connor, S. (1995) Carpenter's Gap Rockshelter 1: 40 000 years of Aboriginal occupation in the Napier Ranges, Kimberley, WA. *Australian Archaeology* 40: 58–59.

Okamoto, G., Okura, T., & Goto, K. (1957) Properties of silica in water. *Geochimica et Cosmochimica Acta* 12: 123–132.

Ollier, C. D. (1969). *Weathering* (revd. edn. 1975). London: Longman.

Oxburgh, R., Drever, J. I., & Sun, Y.-T. (1994) Mechanism of plagioclase dissolution in acid solution at 25ºC. *Geochimica et Cosmochimica Acta* 58: 661–669.

Passemard, E. (1944) La caverne d'Isturitz en pays basque. *Préhistoire* 13: 1–95.

Perez-Gonzalez, T., Jimenez-Lopez, C., & Neal, A. L. (2010) Magnetite biomineralization induced by *Shewanella oneidensis*. *Geochimica et Cosmochimica Acta* 74: 967–980.

Pessis, A.-M. (1999) The chronology and evolution of the prehistoric rock paintings in the Serra da Capivara National Park, Piauí, Brazil. In M. Strecker & P. Bahn (Eds), *Dating and the earliest known rock art*, pp. 41–47. Oxford: Oxbow Books.

Phillips, F. M., Flinsch, M., Elmore, D., & Sharma, P. (1997) Maximum ages of the Côa valley (Portugal) engravings measured with chlorine-36. *Antiquity* 71: 100–104.

Pope, G. A. (2000) Weathering of petroglyphs: direct assessment and implications for dating methods. *Antiquity* 74: 833–843.

Ridges, M., Davidson, I., & Tucker, D. (2000) The organic environment of paintings on rock. In G. K. Ward & C. Tuniz (Eds), *Advances in dating Australian rock-markings*, pp. 61–70. Melbourne: Occasional AURA Publication 10, Australian Rock Art Research Association.

Rimstidt, J. D., & Barnes, H. L. (1980) The kinetics of silica-water reactions. *Geochimica et Cosmochimica Acta* 44: 1683–1699.

Roberts, R. G., Bird, M., Olley, J., Galbraith, R., Lawson, E., Laslett, G., Yoshida, H., Jones, R., Fullagar, R., Jacobsen, G., & Hua, Q. (1998) Optical and radiocarbon dating at Jinmium rock shelter in northern Australia. *Nature* 393: 358–362.

Roberts, R., Walsh, G., Murray, A., Olley, J., Jones, R., Morwood, M., Tuniz, C., Lawson, E., Macphail, M., Bowdery, D., & Naumann, I. (1997) Luminescence dating of rock art and past environments using mud-wasp nests in northern Australia. *Nature* 387: 696–699.

Roberts, R. G., Walsh, G. L., Olley, J. M., Murray, A. S., Macphail, M. K.,

Naumann, I. D., Jones, R., & Morwood, M. J. (2000) Rock-picture chronologies and palaeoenvironmental records from fossil mud-wasp nests: preliminary investigations using optical dating. In G. K. Ward & C. Tuniz (Eds), *Advances in dating Australian rock-markings*, pp. 40–44. Melbourne: Occasional AURA Publication 10, Australian Rock Art Research Association.

Robinson, J. M. (1994) Atmospheric CO_2 and plants. *Nature* 368: 105–106.

Rosenfeld, A. (1985) *Rock art conservation in Australia*. Canberra: Australian Government Publishing Service.

Rosenfeld, A., Horton, D., & Winter, J. (1981) *Early Man in north Queensland*. Canberra: Terra Australis No. 6, Australian National University.

Roy, C. J. (1945) Silica in natural waters. *American Journal of Science* 243: 393–403.

Saint Mathurin, S. de (1975) Reliefs magdaléniens d'Angles-sur-l'Anglin (Vienne). *Antiquités Nationales* 7: 24–31.

Scheffer, F., Meyer, B., & Kalk, E. (1963) Biologische Ursachen der Wüstenlackbildung. *Zeitschrift für Geomorphologie* 7: 112–119.

Schiffer, M. B. (1986) Radiocarbon dating and the 'old wood' problem: the case of the Hohokam chronology. *Journal of Archaeological Science* 13: 13–30.

Schmalz, R. F. (1960) Flint and the patination of flint artifacts. *Proceedings of the Prehistoric Society* 26: 44–49.

Schmid, E. (1958) *Höhlenforschung und Sedimentanalyse*. Schriften des Institutes für Ur- und Frühgeschichte der Schweiz, Volume 13. Basle.

Searle, J. R. (1995) *The construction of social reality*. London: Allen Lane.

Selkirk, P. M., & D. Adamson (1981) The effect of fire on Sydney sandstone. In P. Stanbury (Ed.), *Bushfires and their effect on Australian life and landscape*, pp. 24–31. Sydney: Maclay Museum.

Shaw, T. R. (1992) *History of cave science: the exploration and study of limestone caves, to 1900* (2nd edn). Sydney: Sydney Speleological Society.

Silverman, M. P., & E. F. Muñoz (1970) Fungal attack on rock: solubilization and altered infrared spectra. *Science* 169: 985–987.

Sjöberg, R. (1994) Diagnosis of weathering on rock carving surfaces in northern Bohuslän, southwest Sweden. In D. A. Robinson & R. B. G. Williams (Eds), *Rock weathering and landform evolution*, pp. 223–241. Chichester: John Wiley & Sons.

Smith, D. I. (1978) The micro erosion meter: its application to the weathering of rock surfaces. In C. Pearson (Ed.), *Conservation of rock art*, pp. 44–53. Sydney: Institute for the Conservation of Cultural Material.

Steinbring, J., Danziger, E., & Callaghan, R. (1987) Middle Archaic petroglyphs in northern North America. *Rock Art Research* 4: 3–16.

Suess, E. (1970) Interaction of organic compounds with calcium carbonate. I. Association phenomena and geochemical implications. *Geochimica et Cosmochimica Acta* 34: 157–168.

Syers, J. K., & Iskandar, I. K. (1973) Pedogenetic significance of lichens. In V. Ahmadjian & M. E. Hale (Eds), *The lichens*, pp. 225–248. London: Academic Press.

Taçon, P. S. C., & Garde, M. (2000) Dating beeswax figures on rock art: the view from central Arnhem Land. In G. K. Ward & C. Tuniz (Eds), *Advances in dating Australian rock-markings*, pp. 71–75. Melbourne: Occasional AURA Publication 10, Australian Rock Art Research Association, Inc.

Tang, H., & Gao, Z. (2004) Dating analysis of rock art in the Qinghai-Tibetan Plateau. *Rock Art Research* 21(2): 161–172.

Tarr, W. A. (1917) Origin of chert in the Burlington limestone. *American Journal of Science* 44(4): 409–452.

Tarr, W. A. (1926) The origin of chert and flint. *University of Missouri Studies*, Vol. 1, No. 2.

Tarr, W. A. (1938) Terminology of the chemical siliceous sediments. *Report of the Committee on Sedimentation 1937–1938*, pp. 8–27. National Research Council.

Thackeray, A. I., Thackeray, J. F., Beaumont, P. B., & Vogel, J. C. (1981) Dated rock engravings from Wonderwork Cave, South Africa. *Science* 214: 64–67.

Trendall, A. F. (1964) Examination of rocks with Aboriginal engravings. In W. D. L. Ride & A. Neumann (Eds), *Depuch Island*, pp. 83–88. Perth: Special Publication 2, Western Australian Museum.

Tricart, J. (1969) Étude expérimentale du problème de la gélivation. *Biuletyn Peryglacjanly* 2: 285–318.

Tsai, M.-S., Po, Y. H., & Yang, C.-H. (2006) Formation mechanisms of colloidal silica via sodium silicate. *Journal of Nanoparticle Research* 8(6): 943–949.

Twenhofel, W. H. (1932) *Treatise on sedimentation*. Baltimore: Williams & Wilkins.

Van Tuyl, F. M. (1918) The origin of chert. *American Journal of Science* 45(4): 449–456,

Walker, T. R. (1962) Reversible nature of chert-carbonate replacement in sedimentary rocks. *Bulletin of the Geological Society of America* 73: 237–241.

Walther, J. (1891) Die Denudation der Wüste. *Abhandlungen Mathematisch-Physikalischer Klasse der Königlichen Sächsischen Gesellschaft der Wissenschaften* 16: 112–142.

Washburn, E. W., & Navias, L. (1922a) The products of the calcinations of flint and chalcedony. *Journal of the American Ceramic Society* 5: 565–585.

Washburn, E. W., & Navias, L. (1922b) The relation of chalcedony to the other forms of silica. *Proceedings of the National Academy of Sciences* 8: 1–5.

Watchman, A. (1990) A summary of occurrences of oxalate-rich crusts in Australia. *Rock Art Research* 7: 44–50.

Watchman, A. (1992a) Composition, formation and age of some Australian silica skins. *Australian Aboriginal Studies* 1992(1): 61–66.

Watchman, A. (1992b) Investigating the cation-ratio calibration curve: evidence from South Australia. *Rock Art Research* 9: 106–110.

Watchman, A. (1996) A review of the theory and assumptions in the AMS dating of the Foz Côa petroglyphs, Portugal. *Rock Art Research* 13: 21–30.

Watchman, A. (1998) Composition and source of dust on Split Rock paintings, Australia. *Rock Art Research* 15: 36–40.

Watchman, A. (2000) Micro-excavation and laser extraction methods for dating carbon in silica skins and oxalate crusts. In G. K. Ward & C. Tuniz (Eds), *Advances in dating Australian rock-markings*, pp. 35–39. Melbourne: Occasional AURA Publication 10, Australian Rock Art Research Association.

Watchman, A. L., Walsh, G. L., Morwood, M. J., & Tuniz, C. (1997) AMS radiocarbon age estimates for early rock paintings in the Kimberley, N.W. Australia: preliminary results. *Rock Art Research* 14: 18–26.

Watson, J. A. L., & Flood, J. M. (1987) Termite and wasp damage to Australian rock art. *Rock Art Research* 4: 17–26.

Welch, D. (1995) Beeswax rock art in the Kimberley. *Rock Art Research* 12: 23–28.

Weymouth, J. H., & Williamson, W. O. (1951) Some physical properties of raw and calcined flint. *Mineralogical Magazine* 29: 573–593.

White, D. E., Brannock, W. W., & Murata, K. J. (1956) Silica in hot-spring waters. *Geochimica et Cosmochimica Acta* 10: 27–59.

Williamson, M. A., & Rimstidt, J. D. (1994) The kinetics and electrochemical rate-determining step of aqueous pyrite oxidation. *Geochimica et Cosmochimica Acta* 58: 5443–5454.

Williams-Thorpe, O., Jenkins, D. G., Jenkins, J., & Watson, J. S. (1995) Chlorine-36 dating and the bluestones of Stonehenge. *Antiquity* 69: 1019–1020.

Winkler, E. M. (1973) *Stone: properties, durability in man's environment*. Springer Verlag, New York.

Zeuner, F. E. (1960) Discussion on: Goodwin 1960. In R. F. Heizer & S. F. Cook (Eds), *The application of quantitative methods in archaeology*, pp. 313–324. New York: Viking Fund Publications in Anthropology 28.

In: Weathering: Types, Processes and Effects ISBN 978-1-61324-280-3
Editor: Matthew J. Colon, pp. 69-128

Chapter 2

GEOMORPHIC PROCESSES SINCE THE LATER LAST GLACIAL INDICATED BY THE FORMATION OF BLOCK DEPOSITION FEATURES IN MID-LATITUDE TEMPERATE ZONE

Masayuki Seto
Geosphere Research Institute, Saitama University, Japan

1: PREFACE

1.1. Introduction

A lot of studies on the restoration of the past geomorphic processes on the slopes of the Japanese mountains have been executed mainly for the glacial and periglacial features by the Japanese researchers: Koaze et al. (1974), Ono and Hirakawa (1975), Yanagimachi (1987), Oguchi (1992), Koizumi (1992) and others. Because the slopes in these remaining features are limited to those located at the altitudes that are higher than the glacial forest line, there have been many reports on the subject of the Japan Alps, Kitakami Massif, and high mountains in Hokkaido. On the other hand, a great number of slopes that have not been examined for the past geomorphic processes exist in high zones such as the middle relief mountains of central

Japan where a range of the glacial and periglacial features does not exist. It can be said that the clarification of the formation history and process of such unexamined slopes provides the essential means for investigating totally both the history of geomorphic processes and the history of environmental change in the Japanese Islands.

In order to solve geomorphic problems, it is the most effective method to record the distribution and nature of various slope deposits and estimate their formation processes and periods. However, the slopes of mountains are generally active erosion areas where it is hard to leave reliable evidence including useful information.

Nevertheless, the slope deposits reflecting the past geomorphic processes may remain on the slopes of mountains. Among these slope deposits, the block extending linearly toward the lower part of the slope and having a width of some meters to a few 10 meters and a length of a few 100 meters is sometimes called a "block stream". Such block deposition features exist on the slopes of mountains at relatively lower altitudes and most of them are considered to keep their formation processes unchangeable in the present weather environment (Kubo, 1999). Therefore, the block stream will be able to become a material to reveal the history of slope formation in mountains and the history of environmental change in mountains. The fact that a lot of blocks were produced and might further flow in a certain period is important for investigating object production and mass movement processes on the slopes of mountains.

At the present time, the basic items including the block-stream formation process and period and the environment at the time of formation are not sufficiently clear and even the definition of the words "block stream" is not made perfectly clear. While many case studies in this topic will be reported in future, the geographical features that have been collectively called a "block stream" may be subdivided into multiple features.

Moreover, the term "block stream" used in spite of its unclear formation process may cause misunderstanding that the blocks flowed on the slope of mountain. Thus, this study will use a designation "block deposition feature" as the term not to suggest an origin of the generally called "block stream" that is aggregation of blocks. However, Chapter 2 will take the historically used term "block stream" into consideration and use this term on a timely basis when reviewing prior researches. In addition, the term "block deposits" is used for discussing the stratigraphy of constituents of block deposition features, and this term is also applied timely to individual blocks staying on the slopes of mountains.

1.2. Purposes and Methods of this Study

As described above, the block deposition features are available as the material to estimate the past geomorphic processes on the slopes of mountains. The purposes of this study are (1) to reveal the formation processes of block deposition features and (2) to clarify the geomorphic processes in the middle relief mountains of central Japan from the last glacial period to the postglacial period.

This study uses the following methods to accomplish the purposes. The formationprocesses of block deposition features is investigated with detailed descriptions of their 3D morphology, the shape and properties of constituent blocks, the geomorphic positionsand the circumstances. This study calculates the frequency of freeze-thaw cycles (FTC) on the surface of blocks and discusses the present and past formation environment of block deposition features based on the data of calculations. Further, the formation period of the block deposition features is viewed on the stratigraphic relationship between the slope deposits involving tephras and the block deposits.

The KobugaharaPlateau in the northern part of the Ashio Mountains was selected asthe study area because the slope deposits are stored in relatively normal condition, the block deposition features exist, this area has not been examined completely for the geomorphic processes in the last glacial period, and the meaning of this study can be fully recognized from the viewpoint of the Quaternary research and the geomorphology (Figure 1-1). Chapter 1 explained the meaning, purposes and methods of this study. Chapter 2 will review prior researches on the formation processes and periods of the block deposition features and point out their problems. Chapter 3 will describe the landform and geology of the Kobugahara Plateau in the northern part of the Ashio Mountains that is the observation area of this study, and survey its present geomorphic processes and environment. Chapter 4 will state the shapes and properties of the block deposits and block deposition features (facies, 3D morphology, distribution, existence of water flow at sides and basement of block deposition features, distance between surface of earth and top weathered layer, etc.) and discuss their formation processes. Chapter 5 will examine whether the periglacial processes were concerned in the formation of block deposition features and discuss a role that the former played in the latter. Chapter 6 will mention the stratigraphic relationship between the slope deposits involving tephras and the block deposits and estimate the formation period of the block deposition features. In addition, this chapter will indicate the period of slope instability that active mass-movement occurred on slopes. Based on the discussions in Chapter 4 to Chapter 6, Chapter 7 will unveil the formation processes and periods of the block deposition features and restore the change of geomorphic processes in the said areas from the later last glacial period to the postglacial period.

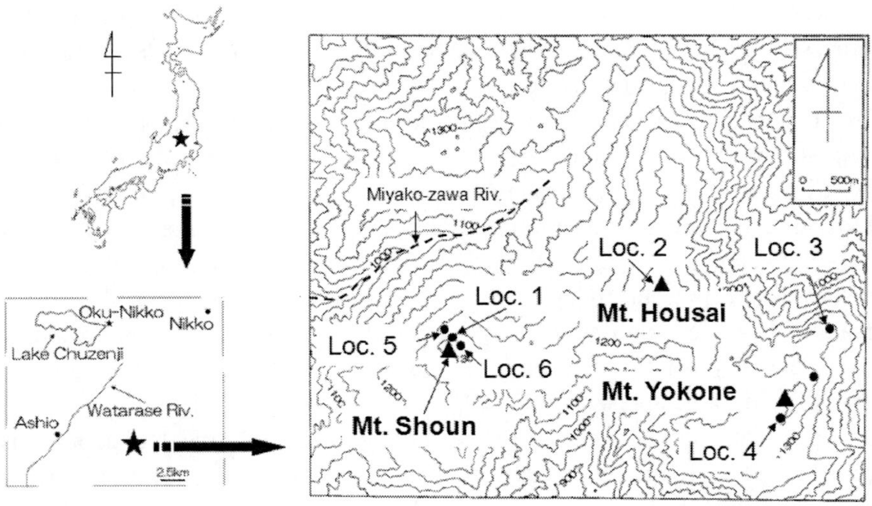

Figure 1.1. Study area and investigation points.

2: Previous Studies on Block Deposition Features

2.1. Discovery of "Block Stream" and its Initial Discussions

In 1845, Charles R. Darwin published the book "Journal of Researches into the Geology and Natural History of the Various Countries Visited during the Voyage of H. M. S.Beagle round the World, etc." in the very early stage of scientific descriptions of the block deposition features. He stated, in a part of East Falkland of this book, that "innumerable large-size angular fragments of Quartzite develop the uncommon 'flow of the stones' on the valley floor at various places of this island", "the remarkable characteristic of this 'flow of the stones' is its low inclination and the 'flow of the stones' inclined at 10 degrees to the horizontal line was detected on the side of a hill, but the inclination of the flat-bottomed valley was just able to be confirmed somewhat clearly" and "this continuous block stream occupies a deep valley and extends to the top of a hill occasionally". The "flow of the stones" stated by C. Darwin is considered to be a landform corresponding to the block stream because of the feature that a lot of roughly angular blocks deposit on the valley bottom with low inclination and the upper part of them extends to the summit of a mountain. Geikie (1894) determined the block stream, which had been mentioned by Darwin, as the landform caused by freezing action. Andersson (1906) imaged that the blocks produced by frost shattering had moved toward the lower part of the slope and formed the block stream due to gelifluction. And then, Joyce (1950) supposed that the blocks that had fallen down a cliff formed the block stream. Washburn (1947) denied that the block stream of the north region of Canada had been formed by viscous flow of fine sediment. Caine (1968) considered that the blocks composing the block stream in Tasmania had moved in the longitudinal direction due to gelifluction and frost creep. Williams (1968) and Small et al. (1970) considered that the block streams existing in Wiltshire country and Dorset country in Britain had been formed after the movement of blocks caused by solifluction in the last glacial period. However, French(2007) supposed that solifluction had brought the said blocks regarded as the ploughing blocks, to the lower part of the slope. As described above, French (2007) considered an origin of the block streams found in Britain to be periglacial mass-movement, but determined there was not sufficient evidence to show that the block streams had been produced in periglacial environment. In other words, the block streams cannot be formed in the present weather environment, and even if they are formed, the fact that the current speed of block stream formation does not sufficiently allow a large quantity of blocks to depositmust be verified. He also stated whether the block streams could be formed or not atthe glacier base was unknown.

Judging from the altitude of mountains and the surrounding landforms, the block deposition features already reported in Japan could not have been formed at the glacier base. There are several examples of the block deposition features in Japan that seem to have been formed in non-periglacial processes during the postglacial period. However, Japan provides many examples of the block deposition features that seem to have been formed in periglacial processes. These examples are perceived as the relic of the block deposition features in the periglacial environment and can be used as the index to restore the past geomorphic processes.

2.2. Problems of Studies on Block Deposition Features

This section reviews conventional discussions of the block production/ mass-movement processes and the formation periods of the block deposition features and points out their problems.

(A) Recognition of Block Deposition Features (Block Streams)

In the block deposits generally called as block streams and their causing microrelief structures, the accumulated blocks are extending linearly toward the lower part of the slope. The accumulated blocks are sometimes distributed over the surface of the slope to cover its large area and are also called felsenmeer or block slope. Meanwhile there are the block deposits such as debris flow deposits, collapse deposits, rock glaciers and glacial deposits that are externally similar to the block stream. Nowadays, it seems that the deposits different from the obvious debris flow deposits and collapse deposits, without enough evidence of the formation mechanism, are called the "block stream". The reason that leads to confusion of terms like the above seems to be that the block deposition features called as the "block streams" show various shape and properties (morphology, case of distribution, and size and shape of constituent blocks) and that the block deposits have also a great variety of formation processes. Therefore, the formation processes and periods of the block deposits called as the "block streams" may be different depending on the case.

(B) Production of Blocks

Because the shape and properties and the formation processes and periods of the block deposition features are not considered to be same in all cases as abovementioned, it is difficult to address them in the general considerations. However, the existing reports clearly indicate that there are a large number of the block deposition features consisting of granite stones so that they were presented by the Japanese researchers: Ichinose (1965), Akagi (1966), Yamakawa (1981), Tanaka et al. (1988), Nomura (1991), Sougaki (1992), Saigusa (1993), Kariya and Saigusa (1993) and Seto (2001). This status suggests that the weathering characteristics of granites (especially, corestone formed as a result of deep weathering) contribute to the block production. Whereas the block deposition features are distributed even in the areas where the rhyolite, andesite and ultramafic rocks exist as the cornerstone so that they were reported by the Japanese researchers: Akagi

(1966), Koaze et al. (1972), Tanaka et al. (1988), Nomura (1991), and Sawagaki (1992).

A point in common between some reports can be found out except the lithology. For example, there are many cases that the block deposition features develop the arborescent plane in the valley with the top of the slope regarded as the upper part so that they were detected by the Japanese researchers: Akagi (1966), Tanaka et al. (1988), Nomura (1991), Tanaka and Nomura (1992), Saigusa (1993), and Seto (2001).

According to the presented reports of the block production processes, there are oneview that the corestones were formed as a result of mechanical weathering or deep weathering on the blocks caused by frost shattering and the other view that the mountain collapse or landslide added the blocks to the block stream (Kubo, 1999). Only a few reports stated where the blocks composing the block slopes or the block deposition features had been generated on the slope in what processes. This indicates that it is hard to answer simply the basic question "Where is the block production source?"

(C) Movement of Blocks

There are various unknown matters in the movement processes of the block deposits composing the block slopes or the block deposition features. It has not been clarified that what process, among mass movement processes caused by periglacial action, played a part especially in the block deposits that the blocks seem to have moved in the periglacial mass movement processes.

Matsuoka (1992) mentioned that it was difficult to explain the movement of block in mass movement such as frost creep and solifluction caused by the freezing and thawing actions because moisture movement did not bring about frost heave in the sand layer and gravel layer that never included fine sediment smaller than silt In addition, he casted doubt on both the talus creep proposed as the process to move the blocks and the estimation that the fine sediment packed between the blocks was moved by frostcreep or solifluction and further, he indicated the necessity of observations and experiments on them.

As described above, it is still difficult to explain in detail that the blocks moved in what processes among the periglacial mass movement processes. However, it is considered that most of the block slopes and the block deposition features that have been already reported fail to conduct their development processes at present (Kubo, 1999). As a result, the periglacial mass movement processes, which were surely more active mass movement processes in the past weather environment different from the present, perhaps have attracted attention.

On the other hand, the movement of blocks in non-periglacial mass movement processes may have been caused by the mountain landslide or collapse and the rock avalanche. The case studies accounting for the movement of blocks in such non-periglacial mass movement processes are limited to the example that the formation period of the block deposition features proved the postglacial period according to historical materials and the example that the block movement processes were confirmed indirectly or directly.

However, mass-movement such as the mountain landslide or collapse and the rockavalanche may have occurred during the glacial period. The continued periglacialmass movement processes even at thebeginning of postglacial period cannot be deniedunder the local conditions including the direction of slope and the altitude and geologyofmountains.

(D) Formation Periods of Block Slopes and Block Deposition Features

The formation periods of the block slopes and block deposition features that have been already reported are divided into the last glacial period and the postglacial period. Whereas, a few reports discussed concretely the formation periods of them based on the sure historical materials of tephra. Figure 2-1 shows the reported cases that the formation periods were discussed based on the historical materials. The formationperiods of the block slopes and block deposition features are different depending on the cases from about 90,000 years ago until now. However, there have been no reports of the formation processes that were activated through both the last glacial period and the postglacial period. This indicates that the conventional studies roughly divide the formation periods and processes of the block deposition features into two parts: the lastglacial period and the postglacialperiod, but it seems to have the following problems. There is a conventional idea that “the block deposition features were formed in their geomorphic processes different from the present under the past weather conditions because the formation processes cannot be explained in the present activated geomorphic processes”. Some examples showed the view that the block depositionfeatures had been formed in the periglacial processes during the glacial period. It is necessary to collect the historical materials and review the formation periods for the block slopes or the block deposition features that have not the evidence of their formation periods.

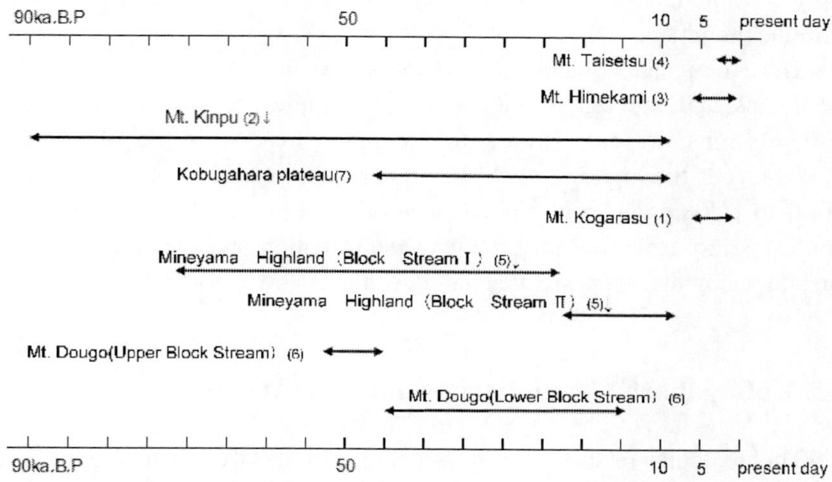

(1) Kariya and Saigusa (1993), (1994)
(2) Shimizu (1983)
(3) Sawaguchi and Hasegawa (1989)
(4) Takahashi (1985)
(5) Tanaka et. al. (1988)
(6) Nomura (1991)
(7) This study

Figure 2-1. Formation periods of block deposition features in Japan.

3: Outline of Obseravation Area

3.1. Landforms of Kobugahara Plateau

The Ashio Mountains are located at the northwest side of the Kanto plains, and the northwest and southwest edge and the east edge of the Ashio Mountains are surrounded by the bottomlands of the Watarase river and the Kinu river, respectively. It seems that the Ashio Mountains were the smaller relief mountains about 3,000,000 yeas ago and were uplifted while tilting gradually in the south-east direction (Kaizuka,1987). The drainage systems in the Ashio Mountains indicate the systematic parallel patterns and most of them are consequent to the landforms. This shows that the Ashio Mountains had the whole landform tilting in the south-east direction before uplifting to form the mountain.

The Kobugahara Plateau is located at the northeast side of the Ashio Mountains and the top of mountains at altitudes ranging from 1,200 to 1,300m forms the easy slopes in the large area, with the Mt. Yokone (1,373.4m) as the highest peak. There is almost no easy slope at altitudes of less than 1,200m, and with the advance of soil erosion, the concavo-convex landforms are generated in the transverse direction and the easy slopes are surrounded by the steep-inclined valley walls.

3.2. Kobugahara Granodioritesand their Weathering

The geologies of the basic rocks composing the Kobugahara Plateau are almost the biotite granodiorite and the hornblende biotite granodiorite that are called the Kobugahara Granodiorites used as a tribute from the late Cretaceous to the early Paleocene (Sudo et al, 1991). The Kobugahara Granodiorites are surrounded by the Jurassic sedimentary complex of the Ashio Terrane consisting of the sandstone, shale, limestone and chert, inclining to the northeast (Sudo et al. 1991).

Generally, the granite stones are often divided into 3 or 4 weathered zones in all. The indexes such as the existence/nonexistence and hardness of the fabric rocks and corestones are used to divide the weathered zones (Ruxton and Berry, 1957; Wilhelmy, 1958; Fumoto ,1974; Kakitani, 1974; Kimiya, 1975). On the basis of these indexes, in 1969, C. D. Ollier specified the weathering zones to the following 6 divisions:

1) Soil
2) Weathering zone without fabric rocks
3) Weathering zone with fabric rocks
4) Weathering zone with fabric rocks, including corestones
5) Weathering zone with fabric rocks, including angular corestones fixed in weatheredlayer
6) Unweathered rocks

There are also the case that the weathering zones were divided in the presence or absence of spring and its color (Shima,1961) and the case that they were divided in the condition and hardness of the minerals (Ohyagi, 1968). Further, Shima (1961) and Kakitani (1974) referred to the velocity of elastic wave in accordance with the divided weathering zones. In the weathering zone classification of the granite stones, there are a lot of cases that the mineral weathering level is generally examined (Ohyagi, 1968; Kakitani, 1974 and Kimiya, 1975).

The weathered layer of the granodiorite on the KobugaharaPlateau includes manycorestones and the remaining fabric rocks are hard to confirm. The appearance of clayey fabric rocks can be detected along the rock joints. These rocks can be easily broken with a hammer and can be also crushed into powder even by hand. There are a lot of feldspars among the minerals According to the weathering zone classification defined by Kakitani et al. (1973) and Kakitani (1974), the weathered layer of the granodiorite on the KobugaharaPlateau is considered to be equivalent to the third chemical weathering layer through the first physical weathering layer, and the velocity of elastic wave in these weathered zones is presumed to be 1.0 to 2.0km/s.

The abovementioned conditions can decide that the top weathered layer of the granodiorite on the KobugaharaPlateau has the same weathering level as "4. Weathering zone with fabric rocks, including corestones" among the divisions specified by C. D. Ollier (1969). The weathering zones 1, 2 and 3 in the weathering zone classification developed by C. D. Ollier (1969) cannot be nearly viewed on the KobugaharaPlateau. It seems that the soil erosion occurred to the weathering zones 1, 2 and 3, and most of them disappeared.

3.3. Climate and Vegetation of Kobugahara Plateau

This section simply describes the climate and vegetation of the Kobugahara Plateau in order to support the examination in the present geomorphic processes, especially the block freezing and thawing actions and the mass-movement processes on the Kobugahara Plateau.

The Oku-Nikko weather station (AMeDAS) is located approx. 15 km north of this area (Figure 1-1) at the almost same altitude of 1,292m as the Kobugahara Plateau, and Figure 3-2 shows the yearly mean values of both the monthly mean temperature and the monthly mean precipitation as data transmitted from this weather station. The monthly mean precipitation is more than 200mm from June to September. During this period, the typhoon invasion or the localized torrential downpour makes the slope erosion more active, so that it seems that the block washing-away processes from the weathered layer or the slope deposits that Chapter 4 deals with are caused more frequently than other periods. However, the monthly mean temperature from December to March is less than zero degrees, and the freeze-thawcycles that Chapter 5deals with occur during this period. Because the monthly mean precipitation of March was 104.4mm, which showed a higher value than December, January and February, and a lot of water flowed on the exposed rocks in comparison with other winter months, these results attract attention in the discussion of the freeze-thawcycles.

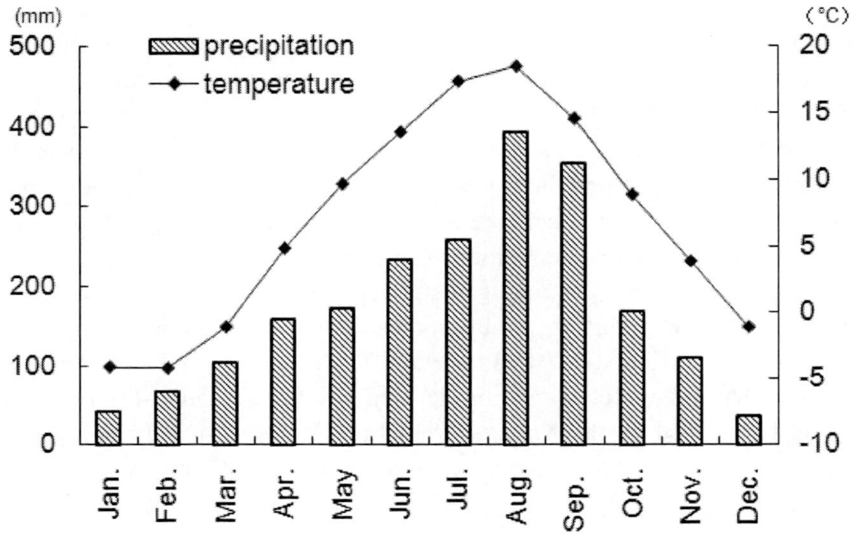

Figure 3-1. Normal monthly temperature and Normal monthly precipitation at Oku-nikko meteorological observatory (1971-2000).

The vegetation in the observation area is a secondary deciduous broad-leaved forest consisting of Betula platyphylla, Sorbus commixta, Malus sieboldii, Quercus mongolica, Enkianthus campanulatus, Rhododendron wadanum and Clethra barbinervis, further including Larix kaempferi and Abies firma. In addition, Sasa nipponica is distributed mainly on the higher solar radiation places such as the upper part of slope. As described in Chapter 4, because the said plants grow from the crevices between the blockdeposits forming the block deposition features, it seems to indicate that the blocks do not move actively at present.

4: Block Deposition Features and Their Formation Processes

4.1. Block Deposits and Block Deposition Features Distributed on Kobugahara Plateau

The block deposition features will go into detail in Section 4-2 and the following. The shape and properties of the block deposition features in this area are generallyrecognized as follows.

There are some cases that the block deposition features are distributed so as to make the waste-filled valley, with the first-order valley at the upper part of the valley wall regarded as the highest part, and its lower part reaches the valley floor. The block deposition features and surrounding areas consisting of Quercus mongolica, Enkianthus campanulatus and Clethra barbinervis are the second-growth forest including the Abies firma, and these plants do not vary both in the block deposition features and on the peripheral slopes, so that it is difficult to clarify the distribution of the block deposition features in an aerial photograph. Various types of blocks forming the block deposition features have the major axis of 50 to 150cm with the non-edge round and angular-edge morphology. The fabric of these blocks without matrix is unknown. The stream of water often exists in the basement of block deposition features. Many lobe microrelief structures are viewed on the block deposition features, but the characteristic that the boulders are converged on the top of them is not detected.

Yamakawa (1981) reported that the block-streamdeposits with the width of a few meters to 10m were found on the bottom of the shallow dish-shaped valley on the easy slopes in this area. These deposits seem to be placed at the almost same area as the block deposition features observed in the Mt. Yokone. However, although the deposits described by Yamakawa (1981) have the characteristic that only the upper half part of the blocks is exposed and their lower half part is buried under the black soil layers or the weathered volcanic ash layers, the block deposition features observed in this study do not have the weathered volcanic ash layers as matrix even when most of the blocks are buried under the ground surface, and there is a slight difference between the characteristics of deposits.

4.2. Formation Processes of Block Deposition Features Viewed From Morphology

4.2.1. Outline of Block Deposition Features in Mt. Yokone

The conventional studies on the block deposition features in this area often focusedon each individual block that was found mainly on the surface of earth or in the slopedeposits. Therefore, it is hard to say that the previous studies have regarded an aggregation of blocks as one colluvial landform, reported the characteristics of the morphology and constituent blocks, and tried fully to discuss the origin of them. Therefore, this study describes the 3D morphology of block deposition features and the shape and properties of block deposits and examine their formation processes.

The observation area is the upper part of valley walls that contact the easy slopes onthe top of the ridges running in south from the top of the Mt. Yokone (Figure 1-1, Loc. 4). The block deposition features are viewed in the first-order valley at the upper part of the slope.

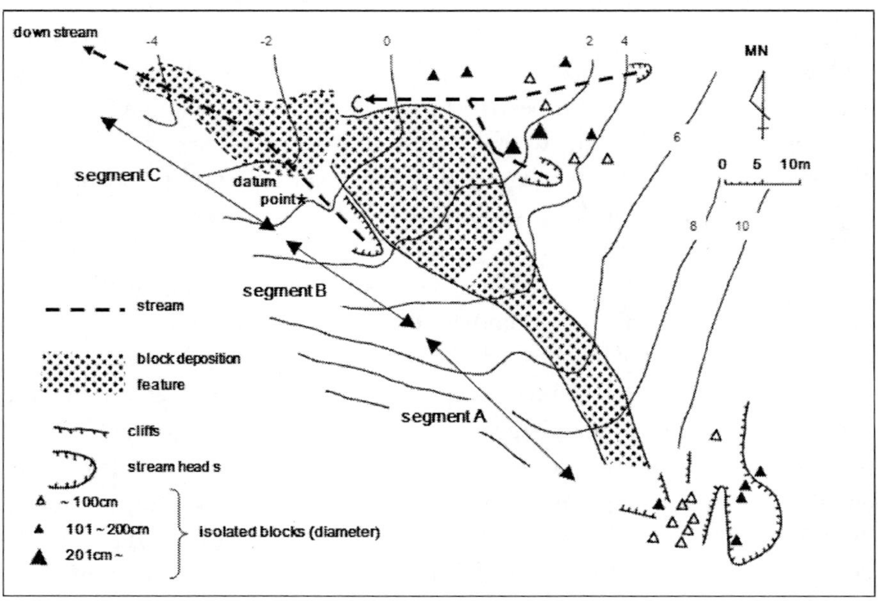

Figure4.1. Surveyed Block deposition feature (Contour interval is 2m. Contour lines show the difference of height with a datum point).

Most of the blocks forming the block deposition features have the major axis of 50 to 150cm, with about 250cm as the maximum. There are the plate-shaped blocks and the non-edge round blocks among the blocks forming the block deposition features. Both types of blocks have the weathered cracks. After depositing, some blocks crack to generate the secondary blocks. The fabric of the block layers without matrix is unknown. Clethra barbinervis is growing from the crevices between the blocks.

As shown in Figure 4-1, there are several low precipices at the upper part of block deposition features. The relative height of low precipices bending at the highest edge is about 100cm and that of others at several edges is about 30 to 50cm. The block deposition features regard the lowest edge under precipice as the top, but the blocks also exist widely on the edge.

The valley heads with the width of 1 to 2m are viewed at the sides of block deposition features where water flows (Figure 4-1). Appearance of a stream of water is that water flows downstream as surface runoff and water goes underground from the crevices between the blocks. The blocks exist here and there around the waterways of surface runoff. The earth and sand, which are the weathering origin of granite stones, depositin the waterways.

4.2.2. Plane Figure of Block Deposition Features

The plane figure and the vertical cross-section figure were measured to investigate the 3D morphology of block deposition features distributed on the Mt. Yokone. The longitudinal line was set along the convex of block deposition features (Figure 4-2). The cross-section line was put along the longitudinal line at intervals of 5m. The results of measurement were shown in the plane view (Figure 4-1), the longitudinal sectional view (Figure 4-3) and the cross-sectional view (Figure 4-4) using the contour line at intervals of 2m.

Figure 4-1 shows the plane figure of block deposition features, the peripheral microreliefstructures and the location of segments to be described latter. The blocks deposit withthe width of some meters from the upper part of block deposition features to the zone ofabout 40m in the longitudinal direction. This zone is called the "Segment A". The width of block deposition features becomes larger in the zone from about 40 to 65m, leading to the max. width of 20m. This zone is called the "Segment B". The blocks are buried under the ground surface from the upper part of block deposition features to the zone of about 40m in the longitudinal direction, but the width of zone where the blocks are distributed is determined to be smaller than Segment B from the viewpoints of the reaction of a pole struck into the ground and the distribution of visible blocks. This zone is called the "Segment C". The shape of the contour lines crossing the block deposition features displays the harmonious valley type with the peripheral valley landforms in the small-width zones (Segments A and C), but it does not create such a valley type in the large-width zone (Segment B).

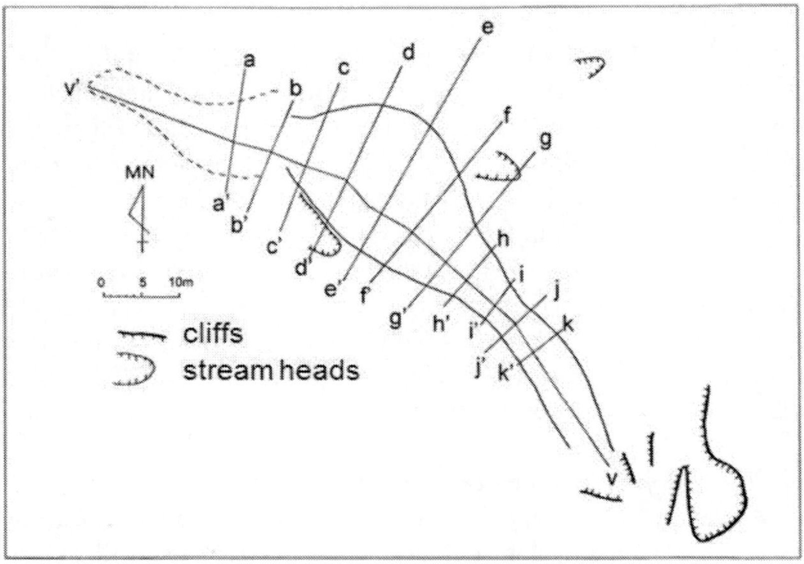

Figure 4-2. Location of topographic sections.

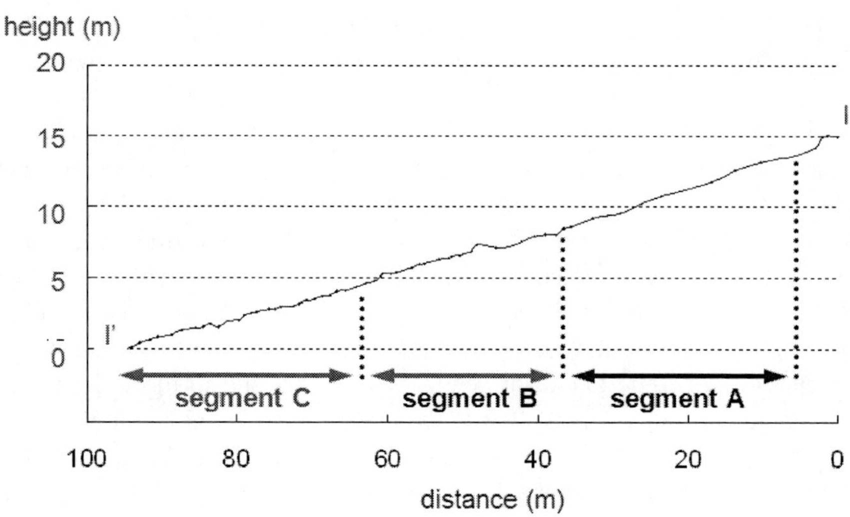

Figure 4-3. A longitudinal section of the block deposition feature.

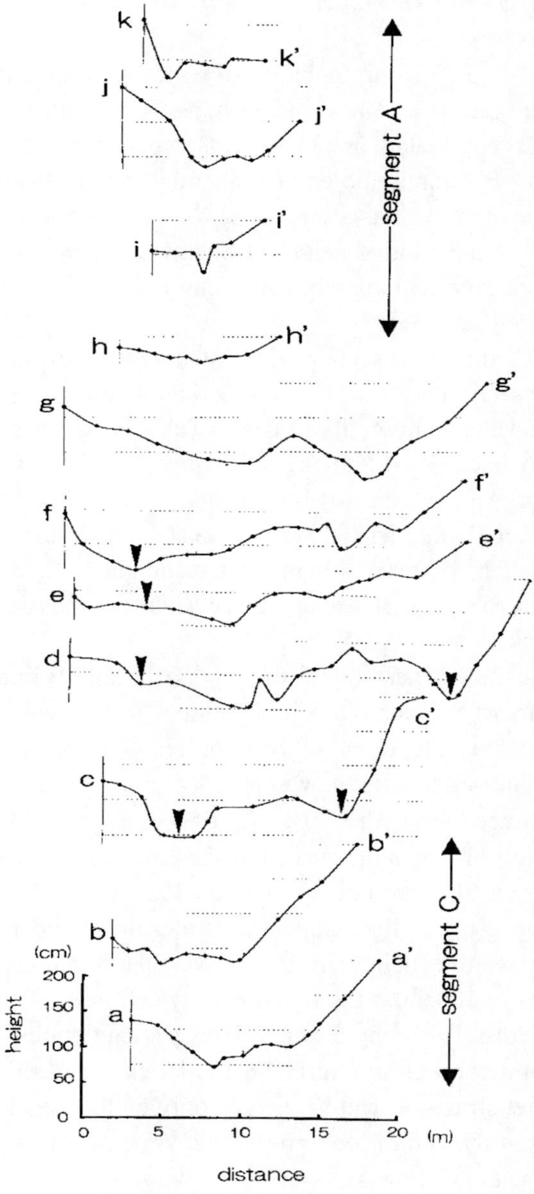

Figure 4-4. Cross sections of the block deposition feature. An arrow and dashed line indicates the location of a stream and the distribution of blocks, respectively. P is locality of Figure 4-2.

4.2.3. Vertical Figure and Cross-Section Figure of Block Deposition Features

The longitudinal line (measuring line 1-1') shows generally the even linear cross section (Figure 4-3). The vertical figure of Segment A is the even linear cross section except the low precipices at the upper part. There are two convexes on Segment B. This is the end of lobemicrorelief structures to be mentioned latter. Segment C has more concavo and convex parts than Segments A and B, but the relative height of Segment C is less than that of Segment B. These concavo and convex parts may be caused by the blocks buried under the ground surface.

Figure 4-4 shows the cross-section view of block deposition features. The number ofcross-sectionlines is 11: (a) – (k). Flour lines of (a), (b), (c) and (d) are placed on Segment A, five lines of (e), (f), (g), (h) and (i) on Segment B, and two lines of (j) and (k) on Segment C. The cross-section view consisting of (a), (b), (c) and (d) lines displays the cross-sectionfigure of Segment A. Although the whole view indicates the cross section of concavo, the convex part of max. 50-cm relative height is detected at the center of (a), (b) and (d) lines. It was confirmed that this convex part was composed of the blocks.

The cross-section view consisting of (e), (f), (g), (h) and (i) lines displays the cross-sectionfigure of Segment B. Although the whole view indicates the cross section of concave, the relief of Segment B is larger than that of Segments A and C. The valley with the width of 1 - 2m runs near the (e) line. This valley can be traced from the (e) to (i) line in Figure 4-4. Water is generated from the top of waterway placed at the sides of block deposition features and flows on the ground surface through the sides of block deposition features up to the (i) line. Underflow water flows under the blocks at the (j) line. The relative height of the convex part on Segment B is measured from 50cm to 100cm at the maximum, which is the highest value between three Segments. This convex part that exists continuously on plane constructs the lobe microrelief structures on the block deposition features. These lobe microrelief structures can be viewed only on the Segment B. The major axis of blocks at the end of lobe microrelief structures is 50 to 150cm as same as the blocks at other zones.

The cross-section view consisting of (j) and (k) lines displays the cross-sectionfigure ofSegment C. Although the whole view indicates the cross section of concavo, the convexpart of less than 50-cm relative height is detected at the center of these lines as well asSegment B. This convex part is composed of the blocks.

4.2.4. Formation Processes of Block Deposition Features Based on Morphology

Because the contour line crossing Segment A or C presents the harmonious valley type with the peripheral first-order valleys, it seems to reflect the basic landform covered with the block deposits. On the other hand, the contour line crossing Segment B displays the relief shape that is not harmonious with the peripheral valley landforms(Figure 4-1). Such a contour line is considered to reflect not the basic landform under theblock deposits but the surface morphology of depositsthemselves, so that it can be estimated that the thickness of layer of the block deposits on Segment B is larger than that on Segments A and C.

Segments A and C seem to have the nature of block migration pathway in consideration of the characteristics of their landforms and the small width of block distribution. On the contrary, Segment B simultaneously presented the nature of deposition zone and the lobe microrelief structures (Seto, 2003). The existence of lobe microrelief structures means that the blocks were not weathered to remain at the present position, but they moved from the upper slope. The same microreliefstructures in local areas were reported by Takahashi (1985), Tanaka et al. (1988), Saigusa (1993), Shimazu (1997). Especially, Kubo and Kobayashi (2001) supposed not rapid mass movement such as debris flow but slow mass movement such as solifluction as the origin of movement processes because there were no general characteristics of debris flow deposits to cause the boulders converge at the end of lobe microrelief structures. A tendency to make the major axis of blocks longer was not detected even at the end of lobe microrelief structures observed in this study.

There is an idea that either non-periglacial action or periglacial action played a part in the movement processes of the blocks (Kubo, 1999). For example, Kariya and Saigusa (1993) supposed that the blocks were washed out from the weathered layer or the collapse deposits in the formation processes caused by non-periglacial action. However, Tanaka et al. (1988) imaged the production and movement of blocks caused by periglacial action.

Whether the abovementioned slow mass movement was of periglacial action is stillunknown. Because the trees grow from the crevices between the blocks, it is hard tosuppose that the blocks are moving at present, and there is a high possibility that theprocesses forming the blockdeposition features are stopped. As a result, it is presumed that this mass movement was activated under the environment different from the present.

Because the blockdeposition features observed in this study is located at the highest part of slope, both the catchment area and the block supply area are small, and the boulders are not converged at the top of lobe, it is difficult to consider a debrisflow as its origin based on total judgment. Slow mass movement is estimated for the movement processes of blocks as well as the examples discussed by Kubo and Kobayashi (2001). It is hard to suppose that the blocks move in the present open work conditions from mechanical viewpoint when taking the size of blocks into consideration. It seems that the blocks moved with matrix.

Kariya and Saigusa (1993) supposed that the corestones were washed out from the weathered layer in the formation processes of the blockdeposition features. There is also a possibility that the block washing-away action occurred because of the accumulated blocks in the valley and the stream of water around the blockdeposition features and under the blocks in the blockdeposition features of the Mt. Yokone observed in this study. However, it is clear that the landforms were not produced only by the block washed out action from the viewpoint of the characteristics of the block depositionfeatures as mentioned above.

The width of block distribution area on Segment C is considered to be small as well asSegment A, but the difference between Segments A and C is that many blocks areburied under the ground surface on Segment C. Therefore, the formation period ofSegment C may be different from that of Segment A.

4.3. Block Deposition Features Distributed on Mt. Shoun and their Origin

4.3.1. Outline of Block DepositsDetected in Mt. Shoun

The tors, the remaining corestones and the block deposition features characterizing the slope in this area are found on the northern slope of Mt. Shoun, and this can be said the typical slope in this study. The following sections describe the block deposits distributed on the northern slope of Mt. Shoun and the block deposition features as an aggregation of the block deposits and discuss the formation processes. The northern slope of Mt. Shoun has the convex break line near the altitude of 1,320m as the highest part where the easy slopes at the mountain top contact the valley walls. The concave break line is viewed near the altitude of 1,305m (Figure 4-5).

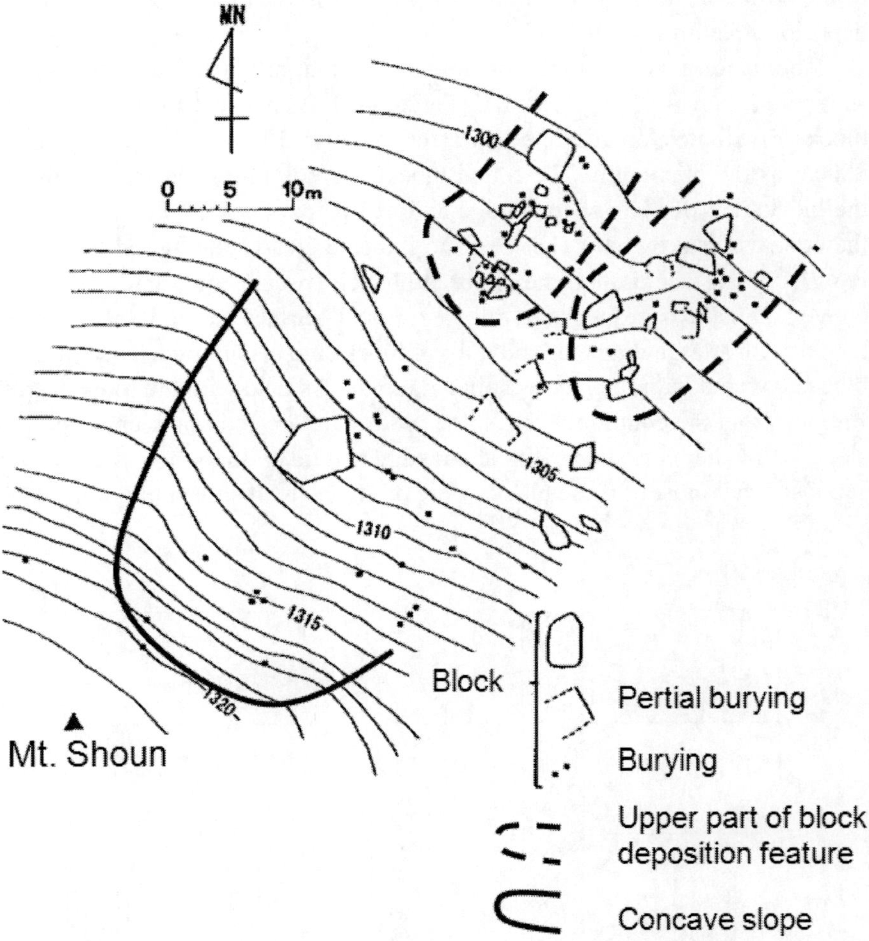

Figure 4-5. Distribution of block deposits on the top of northern side of the Mt. Shoun at Loc. 6. The location is shown Figure 4-7.

The block deposits are distributed in the large area from just under the convex break line on the upper part of slope to the valley floor. There are the tors, which are not shown in Figure 4-5, around the concave break line and at the top of small ridges along the first or second order valleys (Figure 4-1, c). The plate-shaped blocks with relatively sharp edge lie on top of one another in the basement of tors. Almost all of blockdeposits excluding the tors construct the blockdepositionfeatures except the block deposits existing at the upper part of slope. Many individual blocks are distributed at the upper part of the northern slope of Mt. Shoun, which links to the upper part of the block deposition features (Figure 4-5).

The major axis of all measureable blocks that are distributed in the area shown in Figure 4-5 was measured (Figure 4-6). As a result, the major axis of blocks distributed on the upper part of slope is 50 to 150cm, with about 300cm as the maximum. The round blocks are not almost detected, most of the blocks are of plate-shaped edge and lie on topof one another. There are the concavo-convex parts that seem to indicate clearly the buried blocks on the ground surface around the exposed blocks. It seems that many concavo-convex parts exist especially on the ground surface around the exposed blocks with a large diameter and a lot of blocks are also buried. There are the blocks covered with the humus soil and the blocks held with the roots of tree. Judging from the conditions that these blocks are located at the upper part of slope, show the plate shape with sharp edge and lie on topof one another, it is supposed that most of these blocks were produced by disassembly of the tors.

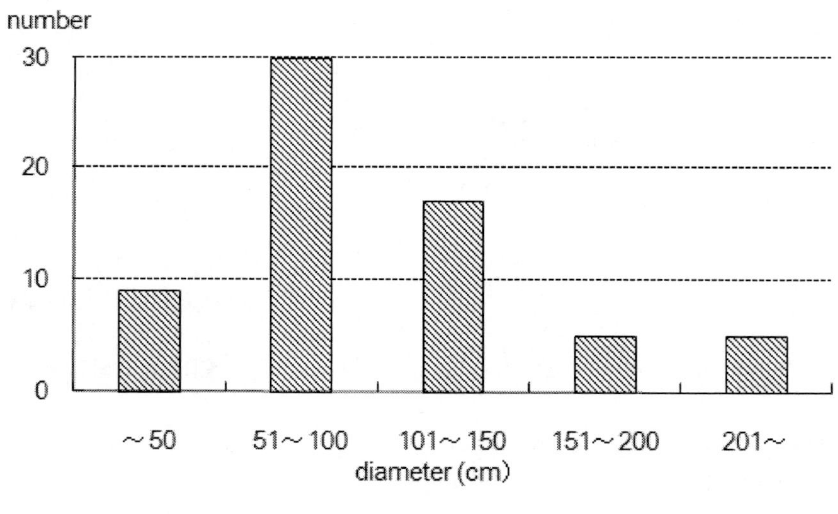

Figure 4-6. Diameter of blocks on top of Mt. Shoun.

Figure 4-5 shows two lines that the blocks are converged linearly on the small-size convex slope. In accordance with the lower slope, these lines increase the density of blocks more, exist on the concavo slope and make the waste-filled valley to form the block deposition features. The two lines that the blocks are converged linearly, as shown in Figure 4-5, exist at the upper part of the block deposition features.

The convex part that is viewed on the slope around the upper part of the blockdepositionfeatures indicates the rise of ground surface caused by the accumulatedblocks. However, the concavo slope is found at the upper part of convex slope and the former slope seems to display the supply source of blocks.

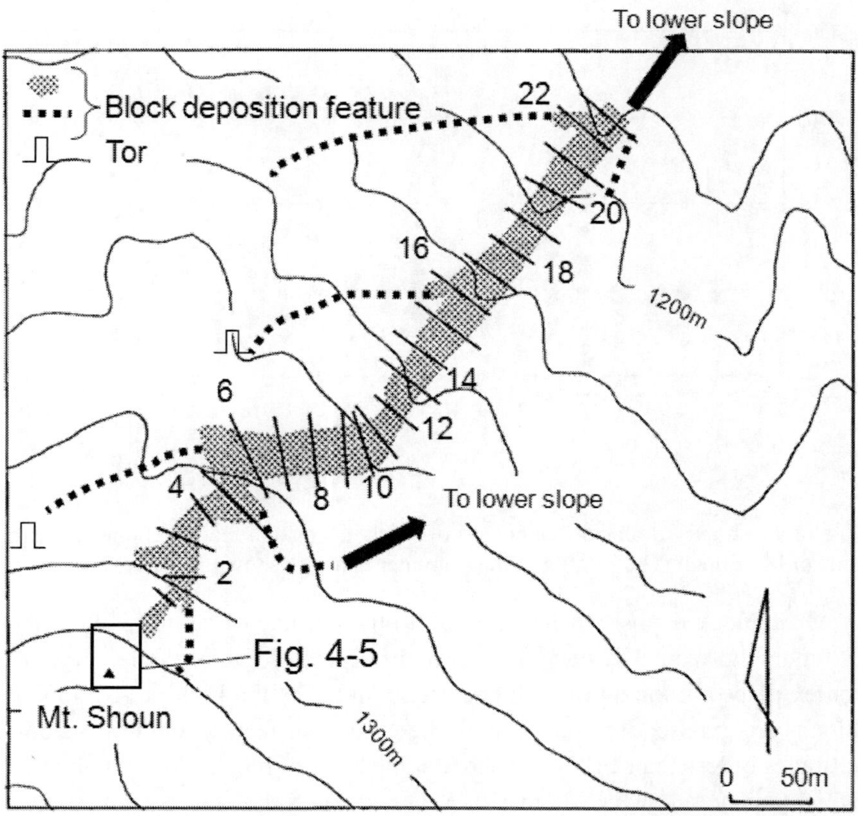

Figure 4-7. Block deposition feature on the northern side of Mt. Shoun.

4.3.2. Block Deposition FeaturesDetected in Mt. Shoun

Figure 4-7 shows the plane view of the blockdepositionfeatures observed in this study. The plane of the blockdepositionfeatures is arborescent and displays the pattern similar to the river system. For the line corresponding to the main stream among the block deposition features shown in this figure, its distribution and plane shape were obtained using the plane-table survey. For the line corresponding to the branch stream, its distribution on the slope was obtained from the distance in the extending direction using the summary survey with clinometers.

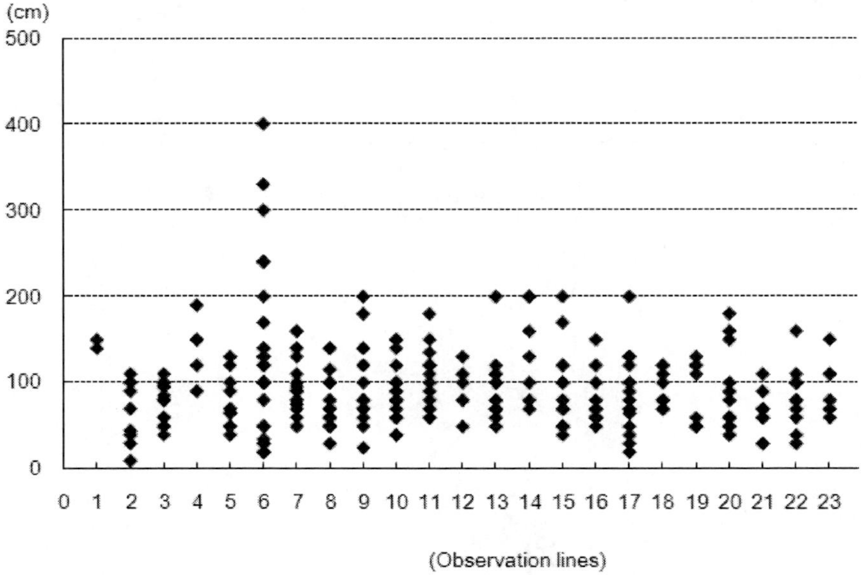

Figure 4-8. b-axis of blocks that consist of block deposition feature on the northern side of Mt. Shoun. The location of measurement line are shown Figure 4-7.

The block deposition features run to the altitude of about 1,220m (about 500m in horizontal distance) toward the lower slope, with the first-order valley placed at the altitude of about 1,305m set as the highest position. The blocks are exposed intermittently to the ground surface on the lower slope at altitudes of less than 1,220m. There are the rows of blocks crossing the ridge, which are not shown in Figure 4-7. The smallest and largest widths of the block deposition features are 5m and more than 20m, respectively.

There are the tors with the width of about 10m and the height of 3 to 4m at the upperpart of b block deposition features. Many plate-shaped blocks were found on the basement of tors. There are no exposed rocks that will be able to become a source to produce the plate-shaped blocks on the slope except the tors. The obvious non-edge round blocks and the sharp blocks are mixed among the blocks composing the block deposition features. Some lobe microrelief structures are found in the block deposition features. These lobe microrelief structures run with the height of about 50cm and the length of 10 to 20m. The characteristic that the boulders are converged on the top of them is not detected.

Concerning the line corresponding to the main stream in the block deposition features shown in Figure 4-7, twenty-three measuring lines crossing the lock deposition features was set at intervals of 20m from the upper part. The middle axis of blocks referring to the measuring lines was measured as much as possible (Figures 4-7 and 4-8). The middle axis of blocks composing the block deposition features is focused to 50 to 150cm. However, there are the cases that the middle axis of blocks reached 200cm in the (6), (9), (13), (14), (15) and (17) measuring lines. Most of these blocks with the large middle axis are of sharp and plate-shaped design. Five blocks with the middle axis of 200 to 400cm were viewed in the (6) measuring line, including the blocks with the middle axis that is remarkably larger than that of other measuring lines. The (6) measuring line shows the area with the largest width among the investigated block deposition features, and other lines presented with the broken lines in Figure 4-7 meet one another at the upper pat of the (6) measuring line. At this meeting point, the blocks with the major axis of about 5m are located to rise at the top of the small ridge, but it is difficult to distinguish these brocks from the tors.

These block deposits have not a matrix. There is the waterway at the side of the block deposition features. The earth and sand as the origin of weathered granite stones deposit in this waterway. Because there are many points where the sound of running water is heard from under the blocks, it is sure that water flows in the basement of the block deposits. These facts suggest that the corestone washing-away action from the weathered layer caused by water flow is a part of the formation processes of the block deposition features. However, the conventional studies have not confirmed the distance between surface of earth and top weatheredlayer around the block deposition features or at its basement and the underground water flowing on the weatheredlayer and have not presented the direct data or reports indicating that the current corestone washing-away action from the weathered layer is at least a part of the formation processes of the block deposition features at the said position. Thus, this study investigated the slope deposits around the blockdepositionfeatures shown in Loc. 5 and checked the distance between surface of earth and topweatheredlayer and the existence of the underground water.

In the investigation, the cross section of landform was measured at right angle in theextending direction of the block deposition features so as to make the measuring linecross the block deposition features. Next, the pit was dug on this measuring line todescribe the existence of the deposits on the surface layer and the blocks under the ground surface (Figure 4-9). The measuring line was set to cross the block deposition features on two lines. The position where the measuring line crosses the block deposition features located between the pits (4) and (5) is near the upper part and have the characteristics similar to Segments A and C mentioned in 4-2-2. On the other hand, the position where the measuring line crosses the block deposition features located between the pits (5) and (6) is under the upper part, is the larger width than the upper part and have the characteristics similar to Segment B.

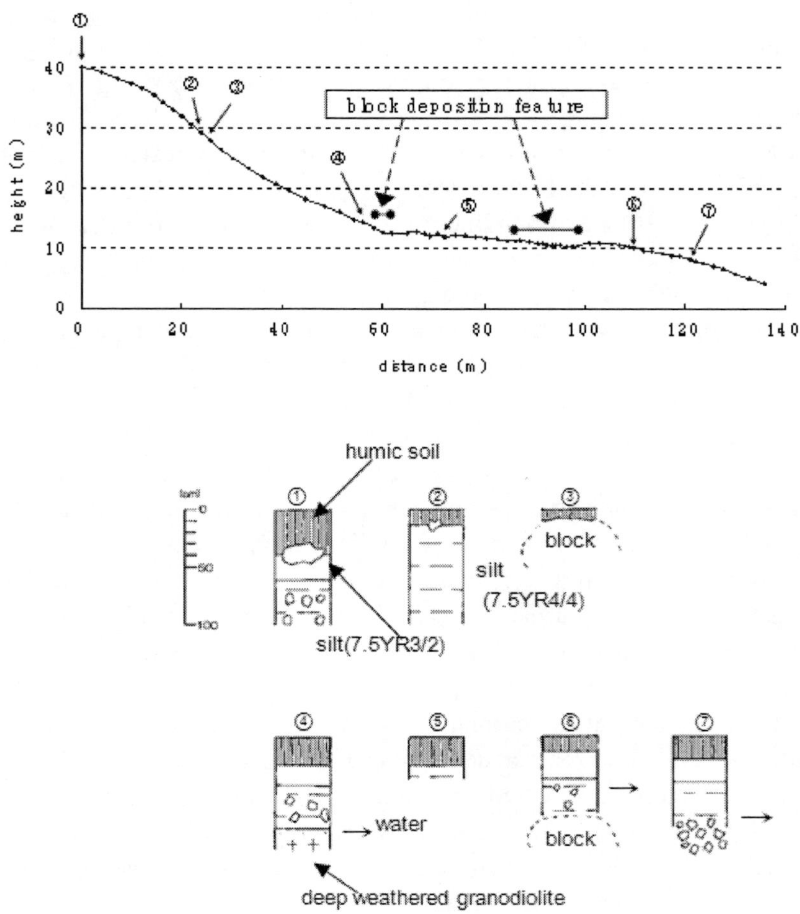

Figure4–9.Cross section and Surface profiles at Loc.5.

The blocks are buried just under the humus soil layer in the pit (3). The weatheredlayer or blocks exist at 65 to 80cm under the ground surface in the pits (4), (6) and (7). The plane breccias with the major axis of a few 10cm are buried at the high density in the lowest pit (7). In the pit (4), water flows from the surface of weathered layer of granodiorite located at about 65cm under the ground surface, and the weathered layer is detected as semipermeable layer. Because water also flows in the pits (5), (6) and (7), it is supposed that the surface of weathered layer is located at the relatively shallow in these areas. The phenomenon above means that water flow washes away the fine sediment and the corestones are exposed to the ground surface as the available processes at least in the investigation positions.

4.3.3. Formation Processes of Block Deposition Features Detected in Mt. Shoun

The tors and the blocks as the origin of tors are distributed in the highest part of slope and the upper part of the block deposition features. Because the blocks detached from the tors construct the upper part of the block deposition features, the tor disassembly can be regarded as one of the formation processes of the block deposition features.

There are many broken blocks at the present positions among the blocks composing the block deposition features. As described in Chapter 5, when taking the frostshattering of tors on the slope into consideration, it can be estimated fully that the block deposits were broken by the frostshattering action.

Appearance of the lobe microrelief structures on the block deposition features seems to display that the blocks were moved on the slope by the slow mass movement on the northern slope of the Mt. Shoun as well as the example of the Mt. Yokone.

At look at the cross section of surface layer around the block deposits shows that the blocks are viewed not only on the ground surface but also under the ground. On this cross section of surface layer, the surface of the weathered layer is at about 1m under the ground surface and water flows here. The fine sediment as the origin of weathered granodiorite deposit in the waterway of ground surface. Based on the above matters, it can be said that on the northern slope of the Mt. Shoun, the blocks are washed away by water flow from the weathered layer and the slope deposits and this action is a part of the formation processes of the block deposition features.

The shape and properties of the block deposition features distributed on the northern slope of the Mt. Shoun were described to discuss the formation processes of them. As a result, it seems that the formation of the block deposition features on the northern slope of the Mt. Shoun was conducted through the processes of the tor disassembly, the corestone washing-away action from the weathered layer caused by water flow and the movement of blocks caused by slow mass movement. However, it is considered that these processes were conducted at a certain position only. For example, it is hard to adopt the corestone washing-away action and the movement of blocks because the tor disassembly seems to be a part of the formation processes of the block deposition features near their upper area, but the water flow and the lobe microrelief structuresare not confirmed.

From the discussion above, it can be said that a single slope process do not play an important role for the formation processes of the block deposition features, but the different multiple processes were activated in the near space at a little different time. In other words, it seems that the processes such as the core stone wash out action from the weathered layer, the slow mass movement (including solifluction) and the tor disassembly caused by the frost shattering action were activated at the different space and they also changed the importance of role played in the formation processes of the block deposition features according to the change of climate.

5: Frost Shattering Action for Exposed Blocks on Slope

5.1. Formation ofBlock Deposition Features and Periglacial Processes

As described in Chapter 2, there have been a lot of reports that the formation processes of the block deposition features are considered to be the same period as the periglacial processes. The formation processes of the block deposition features are estimated on the basis of the block loading caused by solifluction and the tor disassembly caused by the frost shattering action in this area as well as the above (Tabuchi and Hara, 1982; Yamakawa, 1981; Seto, 2003). However, there have not been the positive reports to prove that the movement and production of objects by the periglacial action were concerned in the formation of block deposition features. When discussing the landform processes during the last glacial period, it is very important to recognize the periglacial action in reference with the recognition of climate features such as the glacial and periglacial features.

This chapter examine whether the frost shattering action was concerned in the tordisassembly in the formation processes of the block deposition features regarded as the periglacial processes. For this examination, it needs to clarify the frequency offreeze-thaw cycles for the exposed blocks on the slopes at present.

5.2. Observation Point and Method

5.2.1. Outline of Observation Point and Period

As mentioned in Chapter 4, many block deposits are distributed on the slopes in this area regardless of formation of the block deposition features. The exposed blocks to be observed for the block temperature are desirable to be the blocks or tors with a large diameter that are the production sources of the block deposits. Thus, this study specified the following two points as the observation point.

Point (1) is the tor found at the upper part of block deposition features on the northern slope of the Mt. Shoun (Figure 1-1, Loc. 1, Figure 5-1 and Photo 4-1, a). The size of this tor is approx. 500cm (H) x 80cm (W) x 300cm (D). This tor located at the altitude of about 1,280m is about 80m away from the convex break line to the lower part of the slope where theon-ridge flatland and the hillside slope contact each other (Figure 5-1, D). The slope contacting the basement of tor is always wet because the first-order valley begins from the side of tor. The plate-shaped blocks with the major axis of about 200cm (max. 242cm) and the minor axis of 50cm (max. 78cm) deposited one another at the downward basement of the tor, and these blocks compose the upper part of the block deposition features (Figure 5-1, C). The observation surface at the point (1) where the block temperature was measured is sloping 86° toward N50E. The plants covering over the observation surface were Clethra barbinervis, Malus sieboldii and Quercus mongolica and their branches and leaves prevented solar radiation in observation on October 7, 2003.

Photo 4-1. Block deposits and tor on Kobugahara. A) core stone (in situ), B) block (originated from tor), C) tor and D)angular boulders (originated from weathering rind).

Point (2) is installed in the block on the northern slope of the Mt. Hosai (Figure 1-1, Loc. 1 and Figure 5-2). The major, middle and minor axes of the block are all about 150cm and it looks like a square. This block located at the altitude of about 1,340m is about 40m away from the on-ridge flatland to the lower part of the slope (Figure 5-2, D).

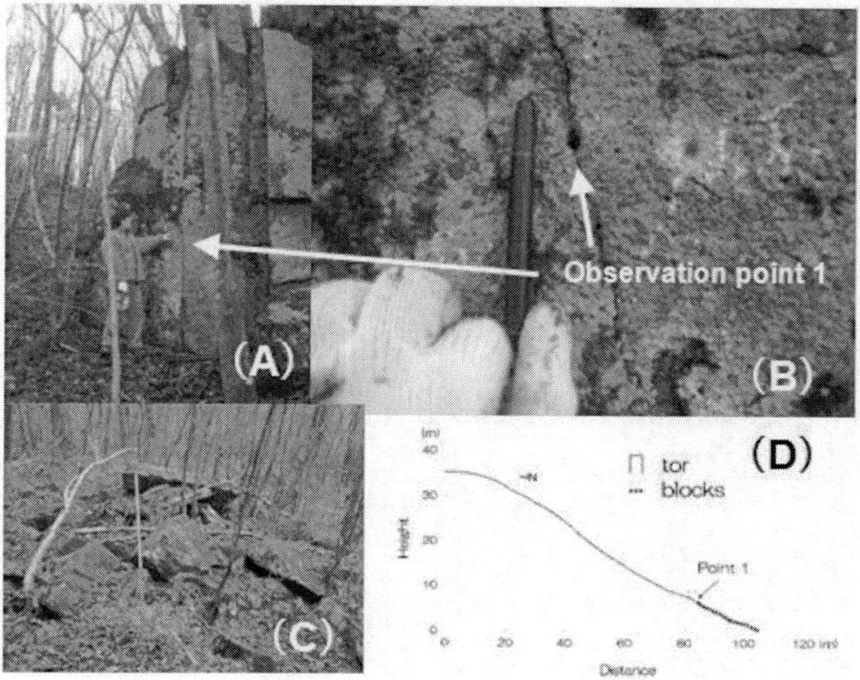

Figure 5-1. Outline of the observation point 1 at Loc. 1.) Whole image of the point 1, B) Close up to the observation hole, C) Blocks around the point 1 and D) A longitudinal section and location of the point 1.

The peripheral fields are drier than the Point (1). The plate-shaped blocks with the major axis of about 300cm (max. 530cm) and the minor axis of 70cm (max. 78cm) are distributed on the peripheral fields as shown in Figure 5-2, C. The observation surface at the point (2) where the block temperature was measured is sloping 88° toward N24E. The plant covering over the observation surface was Larix kaempferi and the point (2)received more solar radiation than the point (1) in observation on October 7, 2003.

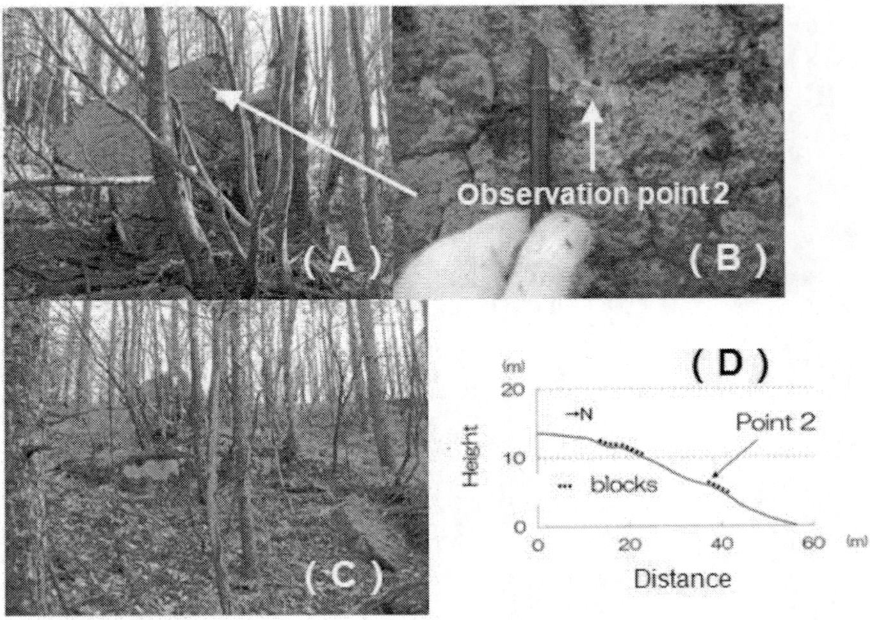

Figure 5-2. Outline of the observation point 2 at Loc. 2. A) Whole image of the point 1, B) Close up to the observation hole , C) Blocks around the point 1 and D) A longitudinal section and location of the point 2.

The total of observation period to use data in this study was 158 days from November 30, 2002 to May 6, 2003. It is necessary to adjust the difference in the measured values between the observation period and the normal year because of the short period of observation in this study. Figure 5-3 displays the yearly changes of the monthly mean temperature for 4 months in winter from January of 1961 to May of 2003 on the basis of data collected from the Oku-Nikko weather station (AMeDAS) that is located approx. 15 km north of this area and also stands at the almost same height above sea level (altitude of 1,292m) as this area. This figure shows that the difference in the monthly mean temperature between December and February is large depending on the year in the yearly changes. Attention shall be paid to this period in the followings. The coldest month from January of 1961 to May of 2003 was February of 1984 and the monthly mean temperature was -8.2 Celsius degree. However, there were a few years including such a cold month and the monthly mean temperature in winter as the coldest periodwas -6 Celsius degree to -4 Celsius degree, changing every year. It was the same in winter from 2002 to 2003, i.e., the monthly mean temperature was -

2.2Celsius degreein December of 2002, -4.8 Celsius degree in January of 2003, and -4.2 Celsius degree in February of 2003. Judging from the above, it is not considered that it was especially cold or warm winter from 2002 to 2003 in comparison with the period of about 40 years from 1961.

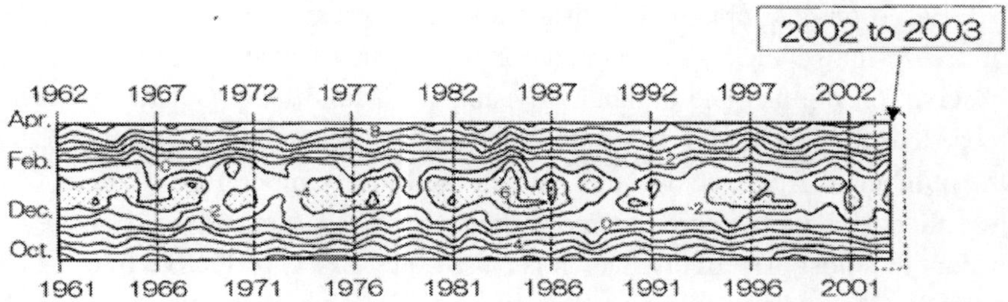

Each vertical line on the diagram shows the change of mean monthly temperature in each season. Horizontal line presents the year to year difference of mean temperature of the same month. Contour interval is 2°C. Dotted area shows under -4°C.

Figure 5-3. Year to year variation of monthly mean temperature during winter four months at Oku-nikko meteorological observatory.

5.2.2. Method for Observation of Freeze-Thaw Cycles

There are the examples of observation of freeze-thaw cycles performed by Matsuoka (1990 and 1994), Shiraiwa (1992 and 1993) and Iwafune (1996). In all cases of them, a hole of 1.5 to 40-cm depth was made on the observation objects, a thermistor was installed in this hole, the block surface temperature was measured at intervals of 1 to 3 hours, and the freeze-thaw cycles were counted. Matsuoka (1990 and 1994) and Shiraiwa (1992 and 1993) packed the silicon and rock dust into the hole.

In this study, a hole of about 5-mm diameter and about 20-mm depth was made in the crack at about 150cm from the ground surface on the north side of the tors and blocks, the silicon was packed with rock dust, the thermistor was fixed in this hole, and the block temperature was measured (Figure 5-1, A and B; Figure 5-2, A and B). The intervals of observation were set 30 minutes. The crack in which the temperature sensor was installed has the width of less than 1mm and the length of about 30cm at all the observation points. The fine sediment was packed into the crack.

Data Logger (Model: Jr. RTR-52) manufactured by T&D Corporation was used for observation. The Data Logger RTR-52 can store data of max.

16,000 pieces and the continuous observation can be performed using batteries for about 6 months. The temperature sensor is the thermistor with the diameter of 2mm, its measurable range is -60 Celsius degree to +155 Celsius degree, and its accuracy is 0.3 Celsius degree on average in the range of -20 Celsius degree to +80 Celsius degree.

Based on data obtained from the above observation method, the number of freeze-thawcycles was counted for the block surface temperature at the observation points. The number of rising from less than 0Celsius degree to more than 0 Celsius degree at the block surface temperature was counted as the number of freeze-thawcycles. When the block temperature changes from less than 2 Celsius degree to more than +2 Celsius degree, frost shattering occurs the most effectively, which is called the "EFTC (effective freeze-thaw cycle)", (Matsuoka, 1990 and Shiraiwa, 1992). This study also counted the number of rising from less than -2 Celsius degree to more than +2Celsius degree at the block surface temperature.

5.3. Block Surface Temperature and Freeze-Thaw Cycle at Observation Points

There were no malfunctions of data logger and thermistor and dislocation of thermistor installed on the observation surface during observation from November 30, 2002 to May 6, 2003. As a result, all data could be collected at Points 1 and 2 without missing value in observation data of block surface temperature.

The results of observation of the block surface temperature are shown in Figures 5-4, 5-5, and 5-6 and Table 5-1. During observation, the simple average value, max. value and min. value of the block surface temperature measured at Point 1 were 0.3 Celsius degree, 18.4Celsius degree (4th of May) and -7.4 Celsius degree (30th of January), respectively. The max. vale and min. value of the block surface temperature of daily average were 15.1Celsius degree (4th of May) and -6.4 Celsius degree (30th of January), respectively. The diurnal amplitude of temperature change on the block surface was max. 8.4Celsius degree (2nd of May). The number of diurnal freeze-thaw cycles was 33 from 16th of December as the first day to the 30th of March as the last day. Among them, the number of FETCs was 6 (Table 5-1).

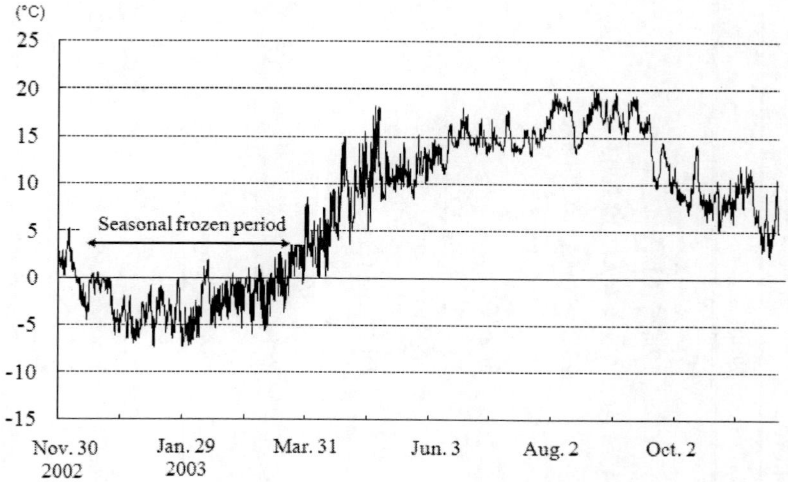

Seto, 2005.

Figure 5-4. Rock surface temperature at point 1.

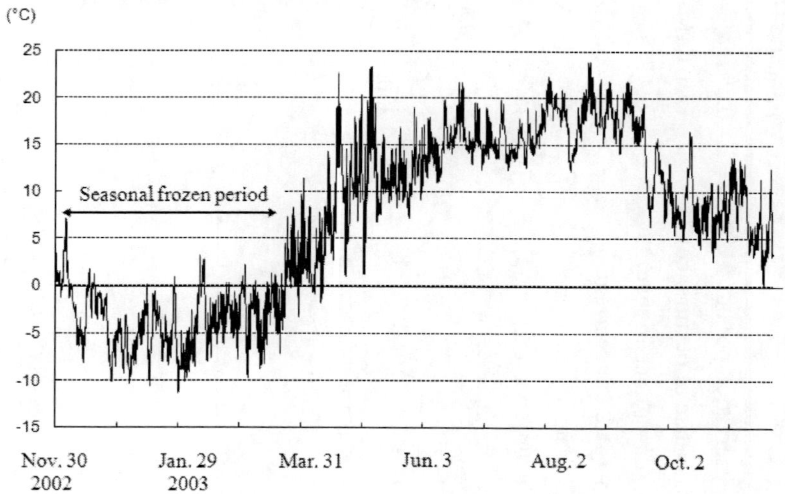

Seto, 2005.

Figure 5-5. Rock surface temperature at point 2.

Table 5-1. Summary of the rock surface temperature observation during the period from November 30, 2002 to May 6, 2003

	Point 1	Point 2
Mean temp. during the observation period	0.3°C	-0.7°C
Diurnal maximum temp. (date of occurrence)	15.1°C (MAY 4 2003)	17.8°C (MAY 4 2003)
Diurnal minimum temp. (date of occurrence)	-6.4 °C (JAN 30 2003)	-9.5 °C (JAN 30 2003)
Extreme maximum temp. (date of occurrence)	18.4°C (MAY 4 2003)	23.3 °C (MAY 4 2003)
Extreme minimum temp. (date of occurrence)	-7.4 °C (JAN 30 2003)	-11.3 °C (JAN 30 2003)
Absolute range during the observation period	25.8°C	34.6°C
Extreme diurnal range (date of occurrence)	8.4°C (MAY 2 2003)	13.6 °C (APR 22 2003 and MAY 1 2003)
Seasonally frozen period (first date-last date)	DEC 16 2002 - MAR 30 2003	DEC 2 2002 - APR 11 2003
The number of times of the freeze thaw cycles	33	30
The number of times of the effective freeze thaw cycles	6	2
Freezing Index	274.1°C · days	438.7°C · days

Seto, 2005.

During observation, the simple average value, max. value and min. value of the block surface temperature measured at Point 2 were -0.7Celsius degree, 23.3 Celsius degree (4th of May) and -11.3 Celsius degree (30th of January), respectively. The max. value and min. value of the block surface temperature of daily average were 17.8 Celsius degree (4th of May) and -9.5Celsius degree (30th of January), respectively. The diurnal amplitude of temperature change on the block surface was max. 13.6 Celsius degree (22nd of April and 1st of May). The number of diurnal freeze-thaw cycles was 30 from 2nd of December as the first day to the 11th of April as the last day. Among them, the number of FETCs was 2 (Table 5-1).

The FETC is found for diurnal freeze-thaw cycles, but the number of the FETCs is less than 10 during the observation period at both observation points. The reason of such a low-frequency of FETC seems to be that the amplitude of the block surface temperature in dailyperiodicity at Point 1 is small. Although the amplitude of that at Point 2 is large, the frequency of FETC seems to be low because the block surface temperature is low through the observation period and is a few times of measured temperature before and after 0 Celsius degrees.

The observation period could be classified into the following three terms: (1) the block surface temperature is almost less than 0 Celsius degree, (2) the block surface temperature often goes up and down from 0 Celsius degree and (3) the block surface temperature is almost more than 0 Celsius degree (Figures 5-4 and 5-5). Among them, the term (1) corresponds to the freezing period for freeze-thawcycles in yearly periodicity. The diurnal freeze-thaw cycles are found in the later March. Therefore, it is made clear that the freeze-thawcycles in yearly and daily periodicities exist at the surface temperature of tors and blocks that were observed in this study.

The diurnal amplitude of temperature change on the block surface at Point 2 was larger than that at Point 1. Its maximum value was 8.4 Celsius degree at Point 1 and 13.6 Celsius degree at Point 2, indicating the difference of about 5 Celsius degree. Because the diurnal amplitude of temperature change on the block surface at Point 2 was also larger than that at Point 1 in this study, it seems that the physical changes such as expansion and contraction are easy to occur in the environment.

The difference in the amplitude of temperature between Points 1 and 2 is considered to be caused by the difference of plants covering over the surface to be measured and the geographical location of observation points. The plants covering over the surface to be measured are Clethra barbinervis, Malus sieboldii and Quercus mongolica at Point 1 and Larix kaempferi at

Point 2. In comparison with the plants at both points, it can be said that even only the density of branches shows that Point 2 is easy to receive more solar radiation than Point 1. The distance from the convex break line where the easy slope and the hillside slope contact each other at the summit of the mountain is 80m at Point 1 and 40m at Point 2, respectively (Figures 5-1, D and 5-2, D). Because both Points 1 and 2 are located on the north slope, it is considered that the slope behind the observation point prevents solar radiation from south. Point 1 located at the lower part of slope is farther away from the convex break line than Point 2 near the ridge, however, the behind slope seems to prevent solar radiation more effectively.

The observation in this study confirmed the freeze-thawcycles in yearly and daily periodicities at both Points 1 and 2, but it was not given data immediately indicating that the frost shattering action was activated in the exposed blocks. The frost shattering action is subject to the lowering level of the block temperature, the frequency of freezing and thawing, the ratio of crevices of blocks, the degree of saturation of water before freezing, and the moisture movement in the blocks in the processes of freezing advance (by Matsuoka in 2001). In order to recognize the frost shattering action from data obtained for the lowering level of the block temperature and the frequency of freezing and thawing in this observation, it needs to collect data such as the ratio of crevices of blocks and the degree of saturation of water. It seems that there is a high possibility that the fine frost shattering action occurred to the exposed block surface since at least the exposed block surface might freeze at Points 1 and 2 in this study and water was supplied from snow, mist and rain. There are many plate-shaped blocks that seemed to be a tor, around the tors and blocks observed in this study, and water is supplied from rain, snow and mist to these tors and blocks even in winter. As the result of measuring the block surface temperature, there were surely the freeze-thawcycles in yearly and daily periodicities. The above contents suggest that the frost shattering action was activated in the disassembly of tors and blocks that were observed in this study. Because the blocks detached from the tor compose the upper part of the block deposition features at Point 1, the disassembly of tor is a part of the formation processes of the block deposition features and the frost shattering action may have occurred to this process.

5.4. Frost Shattering Action Activated in Exposed Blocks on Slopes

The block surface temperature was measured and the number of freeze-thawcycles was counted to examine whether the periglacial processes were concerned in the formation of block deposition features and especially the production/division of blocks. As a result, it proved the following contents:

1) The freeze-thawcycles in daily and yearly periodicities exist at the surface temperature of exposed blocks to be measured.
2) The size and position of tors and blocks often affects the freeze-thawcycles.
3) The diurnal amplitude of temperature change on the block surface at Point 2 was larger than that at Point 1, so that it seems that the physical changes such as expansion and contraction are easy to occur in the environment.
4) The disassembly of tor is a part of the formation processes of the block deposition features in this area. It suggests that the frost shattering action occurred to this process as the result of observation of the block surface temperature.

From the contents above, it indicates that the periglacial processes were activated inthe formation processes of the block deposition features in this area and the slopes inthis area are now placed in the slightly periglacial environment.

6: Period of Slope Instability Viewed From Slope Deposits and Formation Period of Block Deposition Features

6.1. Estimation of Formation Processes and Period of Slope Instability

This chapter describes the stratigraphic relationship between the slope deposits involving tephras and the block deposits forming the block deposition feature and estimate the formation period of the block deposition features. And also, this chapter tries to restore the geomorphic processes in the last glacial period.

The restoration of the geomorphic processes and the estimation of active period of them to be conducted simultaneously are essential to image the geomorphic processes according to the change of climate. When the period of block movement can be specified, such a period is called the "period of slope instability". This means the same idea, made by Tamura and Miura (1968 and 1971), of either the phenomenon that the tephra layer existing on the slope before the tephra fall moved in the secondary action or the slope consisting of the tephra layers is cut additionally, or the condition that based on the breccias layers put between the tephra layers, the period of active movement of objects on the slopes, i.e., the period of slope instability and stability are distinguished.

6.2. Tephras Distributed on Kobugahara Plateau

The tephras creating some key beds fell in this area (Figure 6-1). In order from the lower tephras, there are Akagi-Yunokuchi tephra (Ag-UP: unknown year), Harunayazaki tephra (Hr-HP: 50ka), Akagi-kanuma tephra (Ag-KP: 45 to 50ka), Asama-Itahana brown tephra (As-BP: 20 to 25ka) and Nantai-Imaichi tephra (Nt-I (IS): 14 to 15ka), (by Machida and Arai in 2003). According to Machida and Arai (2003), the stratigraphy of Ag-UP is put between DKP and HR-HP, and the tephra fall year is from 55ka to 50ka.

In order to secure the identification of tephras, the exposures were observed on the mudflow deposits near the fan head of Ohmama alluvia fan (about 30km southwest of this study area) where various types of tephras including the above tephras can be found. Based on the reports of Arai (1956) and Akutsu (1957) , Ag-UP, Hr-HP, Ag-KP As-BP were recognized in this exposure area and their stratification deposits were observed. This study compared and examined the color tone and lithofacies of both the samples obtained in this exposure area and the samples taken out from the deposits on the lower slope and the lithology of breccias included in the pumice layer.

Photo 6-1. Exposure at the Loc. 3. The picture shows central part of Figure 6-1 (b).

The shape and properties of tephras observed on the fan head of Ohmama alluvia fan are mentioned as follows. Ag-UP is the pumice with the grain diameter of about 1cm in light yellow color (2.5Y8/3) and has the chart with the diameter of 2 to 5mm and the breccias of andesite. Hr-HP is the pumice with the grain diameter of 1 to 3cm in ash-white color (2.5Y8/2) and has the breccias of andesite with the grain diameter of 1 to 2cm. Ag-KP is the pumice with the grain diameter of 2 to 4cm in yellow-orange color (10YR8/6) and has the breccias of chart with the grain diameter of about 5mm. As-BP is the pumice with the grain diameter of 5mm to 3cm in yellow-orange color (10YR8/6) and has a lot of sands consisting of andesite.

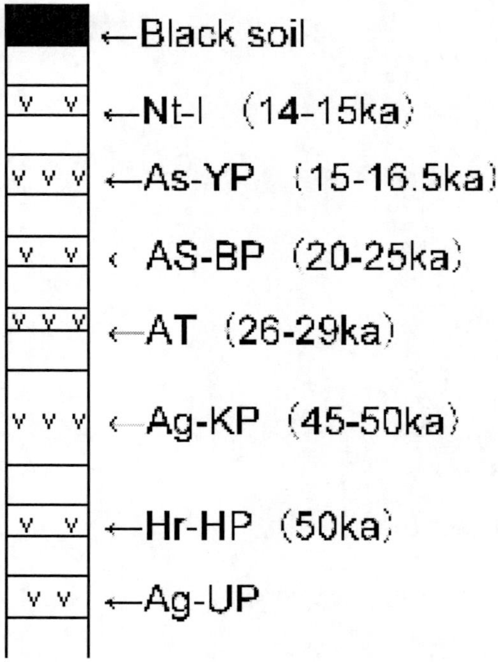

Machida and Arai, 2003.

Figure 6-1. Key Tephras on Kobugahara plateau.

6.3. Slope Deposits

Figures 6-2 and 6-3 show the sketch and the stratigraphic column of the exposure (Fig 1-1, Loc. 3) that appeared, during the woodland path work, on the east slope of ridge running from the Mt. Yokone to the north. Photo 6-1 displays the center of Figure 6-2 (b). In this exposure, the weathered layer of basic rock, the deposit on the lower slope, the buried valley deposits and the deposit on the upper slope divided by the clear unconformity surface are identified from the composition, facies and stratigraphy of the deposits.

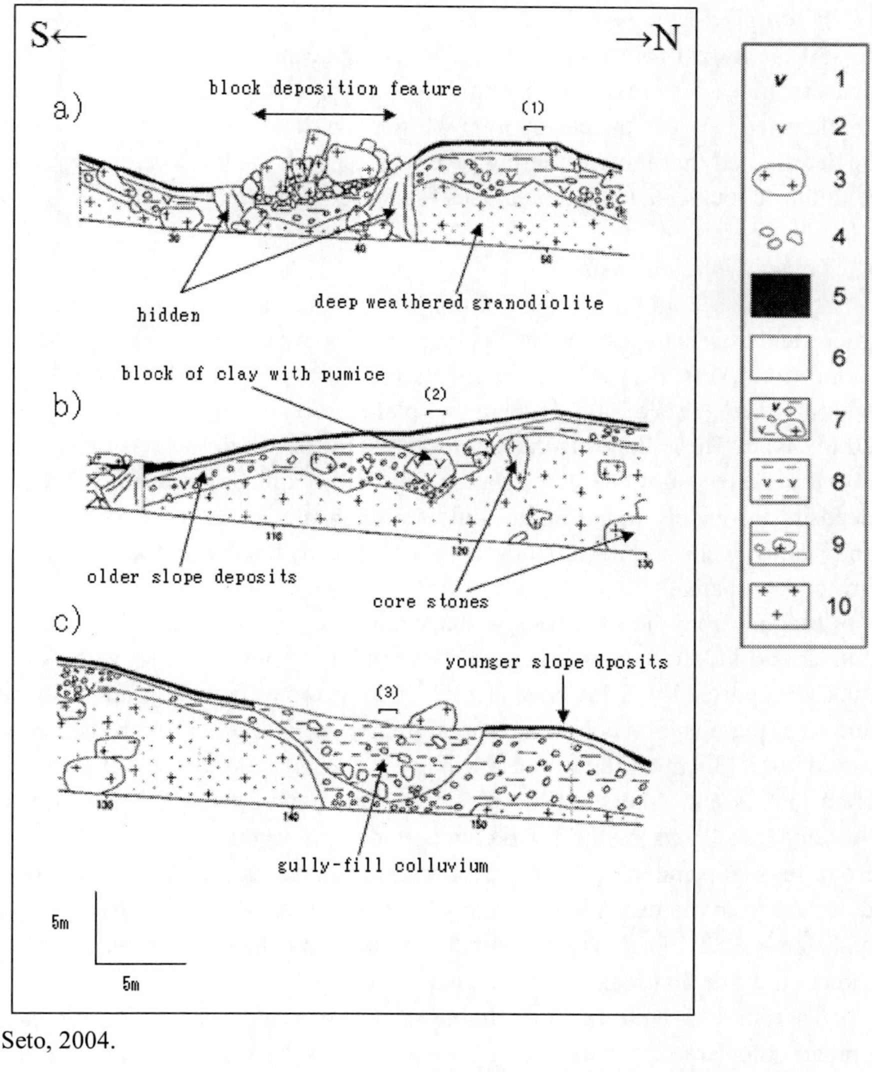

Seto, 2004.

Figure 6-2. Exposure at the Loc. 3. 1) Nt-I, 2) Ag-KP, 3) block of granodiorite, 4) angular boulder, 5) humic soil (7.5YR2/2), 6) silt (7.5YR3/2), 7) silt (7.5YR4/4) with block of granodiorite, angular gravel, Ag-KP and Nt-I, 8) clay (10YR5/8), 9) silt (10YR4/3) with block of granodiorite and angular gravel and 10) weathered granodiorite.

(1) Weathered Layer of Basic Rock

At the lowest part of exposure, there is the deep-weathering granodiorite that can be easily broken by hammer. The onion-like structured core stones are detected between the clayey joints (Figures 6-2, 6-3 and Photo 6-1). It can be determined that the weathering level corresponds to "4. Weathering zone with fabric rocks, includingcorestones" specified by Ollier (1969).

(2) Lower Slope Deposits

The fine sediments as the origin of brown volcanic ash including pumice grains and blocks deposit on the upper part of the weathered layer (Figures 6-2, 6-3 and Photo 6-1). These sediments include the non-edge round blocks with the major axis of 50 to 100cm and plane breccias with the major axis of 20 to 30cm. These deposits displaying non-stratified and poorly-stored have not the facies showing the relationship with laminar waterflow. These deposits are widely found on not only the exposure but also the easy slopes and valley walls at the mountain top and the path up a mountain in the investigation area.

Judging from the stratigraphy, shape and size, the non-edge round blocks correspond to "the boulders having the weathered volcanic ash layers as the matrix" reported by Yamakawa (1981). Judging from the stratigraphy, shape and size, the plane breccias correspond to "the plane breccias with the major axis of 20 to 30cm on the lower volcanic ash layers" reported by Tabuchi and Hara (1982) and "the scalelike plane breccias having the sandy weathered volcanic ash layers as the matrix" reported by Yamakawa (1981). Judging from the shape and size, these breccias seem to be a chip of weathering rind detached from the onion-like structured corestones as pointed out by Tabuchi and Hara (1982). There is no orientation in these blocks and breccias, and the direction of the major axis of them is different.

The pumices with the grain diameter of 1 to 3cm (max. 6cm) in yellow-orange color are distributed on this deposit. This fully clayey pumice grain can be easily crushed by fingers. The clay including high-density pumice grains is lumpy inter-bedded on some of the exposures observed in this study (Figures 6-2 (b) and 6-3 (2)). A lump of clay seems to be an aggregation of weathered pumices because it include the high-density pumice grains. The scoria with the grain diameter of about 3 to 4cm (max. 5cm) in red-brown color exists at the upper part of the deposit on the lower slope. The grain diameter of this red-brown scoria is larger than that of pumice grains in yellow-orange color pre-mentioned. The incomplete clayey scoria can not be crushed by fingers. The red-brown scoria can be easily distinguished from fully clayey pumice grain in yellow-orange color on the basis of color tone and hardness of the grain diameter.

The fully weathered pumice grains distributed widely on the deposits on the lower slope is the pumice with the grain diameter of 1 to 3cm in yellow-orange color (7.5YR7/8) and has the breccias of chart with the grain diameter of about 5mm. After comparing the grain diameter and the color tone of the pumice grain described in section 4-2 with those of this pumice grain and the grain diameter/lithology of breccias, it was determined that the pumice grains distributed widely on the deposits on the lower slope was Ag-KP.

Because the red-brown scoria converges at the top of the deposits on the lower slope,the stratigraphic horizon of scoria is higher than that of Ag-KP. The grain diameter of this scoria is 3 to 4cm and its color is red-brown (5YR4/8) to Black (10YR2/1). As-BP or Nt-I may be available according to horizon, but its color tone is completely different from As-BP in light yellow or white. However, Akutsu (1957) described that Nt-I is in yellow-brown color or deep red-brown color. And also, As-BP is the pumice, but Nt-I is also called the “Imaichi scoria” and is scoria quality (Machida and Arai, 2003). Judging from the above contents, it was determined that the red-brown scoria included in the deposits on the lower slopes was Nt-I.

(3) Buried Valley Deposits

The buried valley deposits bury the valley with 10-m width and 5-m depth by using the deposits on the upper slope and the deposits on the lower slope (Figure 6-2 (3)). The basement does not reach the weathered layer. This deposit consists of the blocks as the matrix of rough yellow-brown fine sediment without including the pumice (Figure 6-2 (c) and Figure 6-3 (3)). The blocks are remarkably weathered granodiorite with the diameter of a few centimeters to 50cm and the non-edge round blocks are mixed with angular blocks. The density of blocks is higher than that of the deposits on the lower slope.

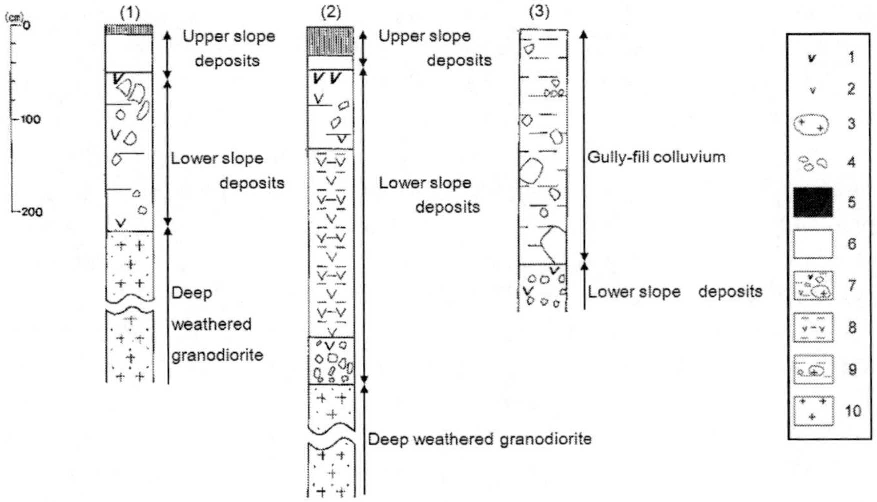

Figure 6-3. Columnar sections of slope deposits at Loc. 3. 1) Nt-I, 2) Ag-KP, 3) block of granodiorite, 4) angular boulder, 5) humic soil (7.5YR2/2), 6) silt (7.5YR3/2), 7) silt (7.5YR4/4) with block of granodiorite, angular gravel, Ag-KP and Nt-I, 8) clay (10YR5/8), 9) silt (10YR4/3) with block of granodiorite and angular gravel and 10) weathered granodiorite.

(4) Upper Slope Deposits

The upper slope deposits consist of the silt layer and the humus soil layer to cover the former (Figures 6-2 and 6-3 and Photo 6-1). The silt deposits on average without layer. Because the black-brown silt layer deposits so as to cover slightly the deposits on the old-period slope and consists of the fine sediment, it may be the origin of volcanic ash. These deposits do not display the disturbed facies and include the objects except silt, they seem to have moved in the short distance even if they moved on the slope. The lowest part of blocks exposing on the ground surface reached the silt layer below the humus soil layer.

(5) Facies of Cross Section of Block Deposition Features

The blocks with the major axis of a few centimeters to max. 200cm deposit in width of about 5m and thickness of 2 to 3m by using the deposits on the lower slope, forming the block deposition features (Figure 6-2 (a)). The deposits on the upper slope are not found on the block deposits. Among the block deposits, there are many angular blocks with the large major axis in comparison with the blocks included in the deposits on the lower slope. The deposit construction is inverse-grading. Water flows on the basement without matrix.

6.4. Formation Processes of Deposits on Lower Slope

The cross section of surface layer in Loc.3 consists of the deposits on the upper slope, the deposits on the lower slope, the buried valley deposits and the weathered layer of basement (Figures 6-2 and 6-3). The blocks and breccias included in the deposits on the lower slope have not orientation and are scattered in the deposits with the fine sediment, which is the origin of volcanic ash, as the matrix. This indicates that all deposits on the lower slope are disturbed. Even if the cross-section figure of the surface of weathered layer is concavo or convex, the deposits on the lower slope deposit so as to cover it. Because the deposits on the lower slope deposit regardless of facies, deposition construction and landform, it seems that they did not deposit in the linear movement of a debris flow, but they deposited by slow mass movement on the surface.

6.5. Recognition of Period of Slope Instability and Formation Period of Block Deposition Features

The deposits on the lower slope include the tephra to be the index of year and are approved widely in this area. Therefore, the formation period of the block deposition features is essential to examine the period of slope instability.

Ag-KP is scattered in the deposits on the lower slope and Nt-I is included at its upper part. Nt-I is different from Ag-KP scattered in a certain width in the vertical direction and is found completely at the upper part of the deposits on the lower slope. Therefore, it is considered that the slope was relatively stable at the time of Nt-I fall. As a result, the period of the disturbed deposits on the lower slope began at the time of Ag-KP fall at the latest and terminated at the time of Nt-I fall or just after that. Because Ag-KP is scattered in the deposits on the lower slope, it clearly indicate that this period of slope instability continued at least after Ag-KP fall. Tabuchi and Hara (1982) pointed out that the blocks and breccias as the origin of the corestones moved on the slope through mass movement and mentioned that only the breccias moved on the slope after the Ag-KP fall. According to Tabuchi and

Hara (1982), the deposits moved by mass movement do not include Ag-KP which makes the layer just over this deposit. In this point, it is different from the deposits on the lower slope where Ag-KP is scattered. Because the exposure observed in this study is close to other exposure observed by Tabuchi and Hara (1982), (about 500m in the linear distance) and Ag-KP is included in the deposit without layer even in the stratigraphic column.

Since the block deposits viewed on the exposure in Loc. 3 deposit irregularly at the upper part of the deposits on the lower slope and form the block deposition features, so that the block deposits is considered to have moved on the slope. The period of starting to move the blocks seems to be on and after the end of last glacial period before and after Nt-I fall because Nt-I is not approved just under the block deposits among the deposits on the lower slope.

There is an interval between the period of slope instability and the period of starting to move the blocks. Although there is the possibility that a part of the formation processes overlaps between the deposits on the lower slope and the block deposition features in relation to the stratigraphy, the formation processes of the block deposition features is more recent.

The reason that the deposits on the upper slope is not found on the block deposits creating the block deposition features is considered that (1) the deposition period of the block deposits is earlier than that of the deposits on the upper slope or (2) the fine deposits on the upper slope could not deposit on the block deposition features where water flows on the basement with matrix free. There is no data or research to determine the condition of time in (1) or the condition of location in (2) as the cause that the deposits on the upper slope do not deposit on the block deposition features.

7: Geomorphic Processes in Kobugahara Plateau from Last Glacial Period to Postglacial Period

7.1. Formation Processes of Block Deposition Features

The surface of the weathered layer exists at about 1m under the ground surface in the block deposition features of the north slope the Mt. Shoun (Figure 1-1, Loc.5) and this place is the semipermeable layer on which the underground water flows. In the place water flows on the surface of ground, the fine sediment as the origin of weathered granodiorite deposits in the passage. Water flows on the basement of block deposition features in the Mt Yokone (Figure 1-1, Loc. 4) as well as the case of the Mt. Shoun and the weathered granodiorite deposits in the side watercourse. The buried blocks are detected around the block deposition features on the north slope of the Mt. Shoun and the deposits on the lower slope include the blocks composing the block deposition features and the similar blocks in size and shape.

Based on the result of observation, it is supposed that water flow washed away the fine sediment from the weathered layer or the slope deposits and then the blocks were washed away as well. This process is one of the formation processes of the block deposition features. However, the density of the blocks in the weathered layer or the

blocks on the lower slope is low to create the block deposition features by only washing-away action as far as the exposure in Loc. 3 is observed. As Seto (2003) pointed out, it is considered that the block washing-away process only is not the formation processes of the block deposition features.

Next, attention was paid to the morphological characteristics and constituents of blockdeposition features by taking as an example the block deposition features near the Mt. Yokone in order to examine the formation processes of them. The block deposition features, based on the characteristics of plane view,, are divided into the following three segments: A, B and C. Segments A and C have smaller width than Segment B and their contour lines are harmonious with the peripheral valley landforms. On the contrary, Segment B has larger width than Segments A and C and its contour line is unharmonious with the peripheral valley landforms. The lobe microrelief structures exist on Segment B, but there are no characteristics of debris flow deposits to cause the boulders converge at the end of lobe microrelief structures. The existence of lobe microrelief structures may mean that the blocks are not only washed away but also moved to accumulate on the present position. The morphological characteristics may present that Segments A and C have the nature of block migration pathway and Segment B is the block deposition area.

Totally judging from the above facts, the block deposition features located at the upper part of the slope and the small catchment and block supply area, it is hard to image the debris flow as the block movement process and estimate the slow mass movement. Whether this slow mass movement was of periglacial action is still unknown. However, because the plants such as Clethra barbinervis grow fully from the crevices between the blocks and this process seems to be unavailable, it is presumed that this mass movement was activated under the environment different from the present.

Next, in order to examine whether the periglacial processes were concerned in the formation of block deposition features, attention was paid to the tor disassembly caused by frost shattering action, the temperature sensor was installed on both the tor and the blocks detached from the tor, and the block surface temperature was measured. As a result, it proved the following contents:

1) The freeze-thawcycles in daily and yearly periodicities exist at the surface temperature of exposed blocks to be measured.
2) The size and position of tors and blocks often affects the freeze-thawcycles.
3) The tor disassembly caused by frost shattering action is a part of the formation processes of the block deposition features in this area.

Figure 7-1 selects the proper processes among the formation processes of the block deposition features examined in this study and the formation processes of the block deposition features that have been suggested in other studies. Since Figure 7-1 displays many study cases of the already reported block deposition features consisting of granite stones and this study on the block deposition features consisting of weathered granodiorite, this figure was created using the granite stones as a precondition. The process that the periglacial action may have played a part was described using a broken line. The formation of the block deposition features first needs the production of blocks. Discussion in Chapters 4 and 5 proved that the blocks composing the blockdepositionfeatures were formed as the corestones in the weathered layer and were produced by the tor disassembly. This indicates that deep weathering of the granite stones is indispensable for the block production processes. In Figure 7-1, (1) deepweathering was suggested as the beginning of processes to form the block deposition features.

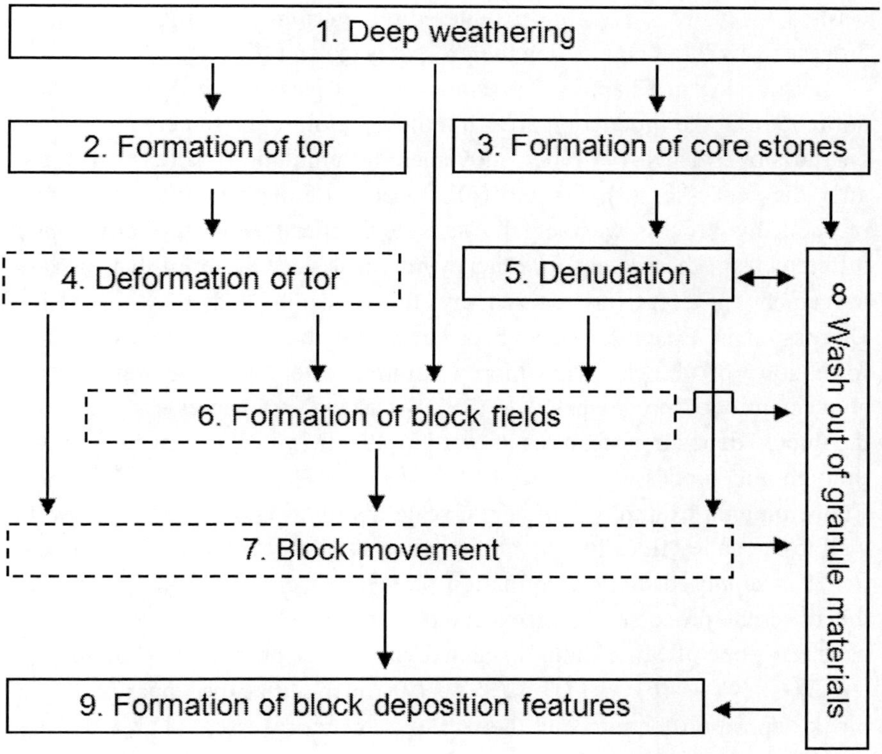

Figure 7-1. Forming processes of granitic block deposition feature. Periglacial phenomena may have played a role in the processes shown in dashed cells.

Second, (2) formation of tor and (3) formation of corestone are carried out as the background of deep weathering of weathered granodiorite. Third, (4) tor disassembly and (5) detaching process are operated. These are mainly the block production processes. It is likely that the whole slope is subjected to the freezing and thawing action from this step depending on the past climate conditions, which forms the block slope as Shimizu (1983) observed. After such a block slope was covered with fine sediment such as volcanic ash, water flow washes away the fine sediment and then the present block deposition features are produced. The weathered layer including many corestones is able to regard as the potential block slope. The block deposition features are generated when water flow washes away the corestones linearly from the weathered layer as well as the example of block slope. Figure 7-1 shows these processes: (6) formation of block slope and (8) removal of fine sediment by water flow. When the corestones in the weathered layer are washed away, it is possible to enter directly from the process (3) to the process (8) without conducting the processes (5) and (6).

Discussion in Chapters 4 presumes that at least a part of the blocks do not remain in the production area, but the slow mass movement moved them on the slope. Thus, (7) block movement/accumulation process was added after the processes (4), (5) and (6). Because of the inactive movement at present, this process was actually activated under the weather environment different from the present. In other words, it is very doubtful that the block movement process was caused by the freezing and thawing actions. Discussion in Chapters 4 and 5 presents that the granite stones reach (9) formation of block deposition features after passing through the abovementioned processes (1) – (8). Discussion in Chapter 4 suggests that the block deposition features viewed in the Mt. Yokone were produced through the processes (1) → (3) → (5) → (7) → (8) → (9) in order. Concerning an example of the great scale distribution from the top of valley wall to the valley floor found on the northern slope of Mt. Shoun, however,

it is supposed that the formation processes were not same in series and the different processes were executed in space. As to an example of the northern slope of Mt. Shoun, it seemed that the tor origination processes ((1) → (2) → (4) → (6) → (7) → (8) → (9)) operated at the upper part of the block deposition features and the washing-away processes ((1) → (3) → (5) → (6) → (7) → (8) → (9)) operated on the lower slope. Most of reports on the block stream in the past explained the whole block deposition feature in one process only, but as revealed from the examples of the Mt. Yokone and the northern slope of Mt. Shoun, it seems to be more normal that the whole block deposition feature was not formed in a certain process, but the different processes were executed in space.

The reason that the processes are different in space seems to be that the conditionsconcerned in the geomorphic processes spatially differ from one another. The conditions to specify the geomorphic processes to be valid at site are considered the following two types: (1) the conditions to affect the whole area and (2) the conditions to be different in the slope or spot (Table 7-1). In case of (1), there are the geology and the weather conditions such as temperature and precipitation. In case of (2), there are the geomorphic positions (ridge, valley floor, upper/lower part of slope), weathered level, catchment area, slope direction, existence of water flow and thickness of slope deposits.

Table 7-1. Effective conditions for geomorphic processes

Wide area conditions	Local area conditions
Latitude, Altitude	Location in the slope
Precipitation,	Catchment area
Air temperature	Water flow
Geology	Weathering
	Slope direction,
	Inclination

7.2. Period of Slope Instability and Formation Period of Block Deposition Features

The stratigraphy of the slope deposits including tephra and the block deposits was clarified to prove the active period of formation processes of the block deposition features described in Section 7-1. The slope deposits in this area are divided into the block deposit, the buried valley deposit, the deposit on the upper slope, and the deposit on the lower slope. The deposit on the upper slope consists of the humus soil layer and silt layer. The deposit on the lower slope includes the block with the silt as a matrix, the breccias (the fragment of weathering rind as the major) and the pumice grain (Ag-KP: c. 45 to 50ka, Nt-I: c. 14 to 15ka). Ag-KP is scattered in the deposits, but the Nt-I isconverged on the surface of deposits on the lower slope. The buried valley deposit consists of the blocks with the silt as a matrix and buries the valley using the deposits on the lower slope. The weathered layer of the basement includes a lot of corestones. The facies of the deposits on the lower slope indicate being disturbed and deposit regardless of the landform. This seems that the deposits on the lower slope were formed by not the linear movement of a debris flow but the slow mass movement on the surface. There is no data or research to determine whether this slow mass movement was of periglacial action. The period of the disturbed deposits on the lower slope was from Ag-KP fall to Nt-I fall. This means that the period of slope instability began at latest about 45,000 years ago and terminated about 14,000 years ago.

The block deposits forming the block deposition features deposit using the upper partof deposits on the lower slope. This suggests that the processes (4) → (7), (6) → (7)and (5) → (7), (block movement/accumulation), as shown in Figure 7-1, started at the lastperiod of slope instability about 14,000 years ago.

7.3. Geomorphic Processes after Later Last Glacial Period

From the synthetic judgment of the new knowledge obtained in this study and the appropriate facts presented by Tabuchi and Hara (1978 and 1982) and Yamakawa (1981), the geomorphic processes in this area after the later last glacial period are considered as described below.

The corestones and tors were formed as a result of deep weathering of the granodiorite, which is a basic rock, in the Kobugahara Plateau before the last glacial period. During about 50 to 14ka after that, the periglacial mass movement produced the fine slope deposits and moved some of the breccias and corestones on the slope. After the last period of slope instability, the slow mass movement moved the brocks on the slope along the valley and further, the block deposition features were formed while being affected by the blocks supplied from the tors and the matrix washing-away action. As abovementioned, there is a time difference between the period of slope instability that the fine slope deposits were produced and the formation period of the block deposition features without matrix, and the latter period is relatively more recent than the former period. This result presents the contents different from both the existing studies supposing that the formation of the block deposition features was conducted by the periglacial action and the existing studies supposing that it was conducted in the non-periglacial processes during the postglacial period. The different contents are that the active periglacial mass movement during the coldest term of theglacial period contributed to the production of the relatively fine slope deposits, but made a little contribution to the formation of the block deposition features. However, not only the water flow action but also the slow mass movements were at least concerned in the formation of the block deposition features. There is the possibility that these characteristics were integrated by the weather conditions during the boundary period between the last glacial period and the postglacial period.

The water flow action remarkably increased in the central Japan region after the late glacial period (for example, by Yamaguchi in 1988), but the same periglacial action as in the glacial period continued on the ridge until the former period of Holocene (e.g. Takada,1986). Finally, it created the environment that both the slow mass movement caused by the periglacial action and the matrix washing-away action caused by water flow were conducted simultaneously from the last glacial period to the former period of Holocene, so that the blockdepositionfeatures may have been formed in the research area.

REFERENCES

Akagi, Y. (1966): Fossil periglacial morphology in Chugoku mountains. *Annals of the Tohoku Geographical Association*, 18, 15-22. (in Japanese).

Akutsu, J. (1957): Kanto loam near Utsunomiya city. *Earth Science*, 33-1-11. (in Japanese).

Andersson, J. G. (1906): Solifluction; a component of subaerial denudation. *Journal of Geology,* 14, 91-112.

Arai, F. (1956): Kanto loam in South Western Akagi volcano. *Earth Science,* 28, 23-30. (in Japanese).

Caine, N. (1968): The blockfields of north-eastern Tasmania. *Australia National University Department of Geography Publish*, G6.

Charles Darwin (1845): *Journal of Researches into the Geology and Natural History of the Various Countries visited during the Voyage of H. M. S. Beagle round the World, etc.*

Fumoto, R. (1974): On the Relationship between the Formation of the Small-relief Erosion Surface and the Weathering of Granite. *Geographical Sciences; Chiri-kagaku,* 21, 11-22. (in Japanese).

H. M. French(2007): *The periglacial environment 3rd edn.* Chichester, 458P.

Geikie, J. (1894): *The great ice age*, 3rd *edn.* Stanford, London, 850P.

Higaki, D. (1992): Pseudo Block Stream Formed by Landslide Movement. *Quarterly Journal of Geography,* 44, 258-260. (in Japanese)

Iwafune, M. (1996): Rock Avalanches Occurred in the Snow-melt Season at the karasawa Cirque in Mt. Hotaka-dake, the Japanese Alps. *Journal of Geography,* 105, 569-589. (in Japanese)

Joyce, J. R. F. (1950): Stone runs of the Falkland Islands. *Geological Magazine*, 87, 105-115.

Kaizuka, S. (1987): Quaternary Crustal Movements in Kanto, Japan. *Journal of Geography,* 96, 223-240. (in Japanese)

Kakitani, S. (1974): Weathering of Granite and Topography in Hiroshima Prefecture. *Geographical Sciences*, 21, 1-9. (in Japanese)

Kariya, Y., and Saigusa, S. (1993): The Origin of Block Streams in and around Mt. Kogarasu, Yamanashi Prefecture, Central Japan. *Quarterly Journal of Geography*, 45, 254-265. (in Japanese)

Kimiya, K. (1975): Tensile Strength as a Physical Scale of Weathering in Granitic Rocks. *Journal of the Geological Society of Japan*, 81, 349-364. (in Japansese).

Koizumi, T. (1992): Progress of Research on Periglacial Smooth Slopes in Japan. *Geographical Review of Japan*, Ser. A, 65A, 132-142. (in Japanese).

Koaze, T., Shimizu, F., Iwata, S., and Okazawa, S. (1972): Block Streams in the kanto Mountains. *Annals of the Tohoku Geographical Association*, 24, 50. (in Japanese).

Koaze, T., Sugihara, S., Shimizu, F., Utsunomiya Y., Iwata, S., and Okazawa, S. (1974): Geomorphological Studies of Mt. Shirouma and its Surroundings, Central Japan. *Sundai Shigaku; Sundai Historical Review*, 35, 1-86. (in Japanese).

Kubo, H. (1999): Studies on the Blockstreams in Japan. *Memoirs of Geography, Graduate School, Hosei University*, 6, 25-32. (in Japanese).

Kubo, H. and Kobayashi. Y. (2001) Block streams in Mt. Himekami, northern Kitakami mountains. *Memoirs of Geography, Graduate School, Hosei University*, 8, 19-29. (in Japanese).

Machida, H., and Arai, F. (2003): *Atlas of Tephra in and around Japan.* University of Tokyo Press, 336P. (in Japanese).

Matsuoka, N. (1990) : The rate of bedrock weathering by frost action: Field measurement and a predictive model. *Earth Surface Processes and Landforms*, 15, 73-90.

Matsuoka, N. (1992): Mechanisms of Frost Action and Related Periglacial Landforms. *Geographical Review of Japan*, Ser. A, 65, 56-74. (in Japanese)

Matsuoka, N. (1994) : Diurnal freeze-thaw depth in rock-wall: Field measurements and theoretical consideration. *Earth Surface Processes and Landforms*, 19, 423-435.

Matsuoka, N. (2001) : Microgelivation versus Macrogelivation: Towards Bridging the Gap between Laboratory and Field Frost Weathering. *Permafrost Periglacial Processes*, 12, 299-313.

Nomura, R. (1991): Colluvial Slopes and Block Streams in the Vicinity of Mt. Dogo in the Central Part of the Chugoku Mountains in Western Japan. *Geographical Review of Japan*, Ser. A, 64, 735-758. (in Japanese)

Oguchi, T. (1988): Differences in Landform Development during the Late Glacial and the Past-Glacial Ages among Drainages around the Matsumoto Basin, Central Japan. *Geographical Review of Japan*, Ser. A, 61A, 872-893. (in Japanese).

Oguchi, T. (1992): An Explanation of Broad Occurrence of Extrazonal Periglacial Landforms in the Lowlands of Western Japan and Korea. *Geographical Review of Japan*, Ser. A, 65, 195-212. (in Japanese)

Ollier, C. (1969) : Zones of Weathering. Ollier, C. *Weathering*, Elsevier. 121-126.

Ono, Y., and Hirakawa, K (1975): Glacial and Periglacial Morphogenetic Environments around the Hidaka Range in the Wurm Glacial Age. *Geographical Review of Japan*, 48, 1-26. (in Japanese)

Oyagi, N. (1968): Weathering-Zone Structure and Landslides of the Area of Granitic Rocks in Kamo-Daito, Shimane Prefecture. *Reports of Cooperative Research for Disaster Prevention*, 14, 113-127. (in Japanese)

Ruxton, B. P., and Berry, L. (1957) : The weathering of granite and associated erosional features in Hong Kong. *Bull. Geol. Soc. Amer.*, 68, 1111-1148.

Saigusa, S. (1993): Block streams in the Southern part of the Kanto Mountains, Yamanashi Prefecture, Central Japan. *Quarterly Journal of Geography*, 45, 266-268. (in Japanese)

Saigusa, S., and Kariya, Y. (1994): Radiocarbon Age of Charcoal Wood from the Forming Material of Block Stream in Mount Kogarasu, Yamanashi Prefecture. *Quarterly Journal of Geography*, 46, 173-175. (in Japanese)

Seto, M. (2001): Block deposits in the Kobugahara, Northern Asio Mountains. *Transactions, Japanese Geomorphological Union*, 22, 414. (in Japanese)

Seto, M. (2003): Morphological Characteristics and Formation Processes of the Block Deposition Feature in Kobugahara, Northern Ashio Mountains. *Quarterly Journal of Geography*, 55, 221-229. (in Japanese)

Seto, M. (2004): Period of Slope Instability and Formation of Block Deposits on the Summit Area of the Ashio Mountains, Central Japan. *Geographical Review of Japan*, 77, 209-218. (in Japanese)

Seto, M. (2005): Freeze-Thaw Cycles on Rock Surfaces in a Montane Zone: Field Measurements in Kobugahara, Norhtern Ashio Mountains, Central Japan. *Transactions, Japanese Geomorphological Union*, 26, 65-77. (in Japanese).

Shima, H. (1961): The Relationship between Weathering and Geomorphology in Granitic Regions. *Anniversary Publication dedicated to Prof. Taro Tsujimura, celeblating his 70 th Birthday*, 202-214.

Shimazu, H. (1997): Characteristics of deposits and debris transport in the headwater area of the Akigawa river, Central Japan. *Geographical Reports of Kanazawa University*, 8, 69-80. (in Japanese)

Shimizu, C. (1983): Fossil Periglacial Slopes on the Chichibu Mountains, Central Japn. Geographical Review of Japan, 56, 521-534. (in Japanese)

Shiraiwa, T. (1992) : Freeze-thaw activities and rock breakdown in the Langtang Valley, Nepal Himalaya. *Environmental Science, Hokkaido University*, 15, 1-12.

Shiraiwa, T.(1994) : Glacial fluctuations and cryogenic environments in the Langtang Valley, Nepal Himalaya. *Contribution from the Institute of Low Temperature Science, Ser. A*, 38, 1-98.

Small, R. J., Clark, M. J. and Lewin, J. (1970): The periglacial rock-stream at Clatford Bottom, Marlborogh Downs, Wiltshire. *Proceedings, Geologist's Association*, 81, 87-98.

Sohma, H. Okazawa, S., Iwata, S. (1979): Slow Mass-Movement Processes in an Alpine Region of Mt. Shirouma Dake, the Japan Alps. *Geographical Review of Japan*, 52, 562-579. (in Japanese)

Sudo, S., Makimoto, H., Hata, M., Unozawa, A., Takizawa, F., Sakamot, T., Komazawa,M., and Hiroshima, T. (1991): *Geological Map, 1:200,000, Utsunomiya*. Geological Survey of Japan. (in Japanese)

Tabuchi, H., and Hara, Y. (1982): Age of Periglacial Features at the Top of the Ashio Mountains. *Annals of the Tohoku Geographical Association,* 34, 164. (in Japanese)

Tada, F., and Oya, M., (1970): Block-Stream in South Korea. *Annals of the Tohoku Geographical Association,* 22, 161. (in Japanese)

Takahashi, N. (1985): The Block Stream on the Chubetsudake NW Slope in the Daisetsuzan Mountains. *Annals of the Tohoku Geographical Association,* 37, 16-28. (in Japanese)

Takada, M. (1986): Fossil Perigalcial Smooth Slopes and Fossil Nivation Hollows along the Main Ridge in the Mikuni Mountains, Central Japan .*Geographical Review of Japan, Ser. A,* 59, 729-749. (in Japanese)

Tamura, T., and Miura, O. (1968): The Period of Slope-Instability Observed in Tephra in a Coastal Area from Hachinohe City to Kuji City, Northeastern Japan. *Annals of the Tohoku Geographical Association,* 20, 69-73. (in Japanese)

Tamura, T., and Miura, O. (1971): Slope Deposits in the Northern Margin of Kitakami Massif, Northeast Japan. *Quaternary Research,* 10, 21-30. (in Japanese)

Tanaka, S., Nomura, R., Inoue, S., Tanaka, T., Tsuchida, K., Ogura, H., and Takada, K.(1988): Landforms and Block Streams in the Mineyama Highland at the Central Part of Hyogo Prefecture, West Japan. *Geographical Review of Japan, Ser. A,* 61, 851-871. (in Japanese)

Tanaka, S., and Nomura, R. (1992): The Depositional Slopes and Their Formative Environment in the Eastern Part of Chugoku Mountains Since the Late Pleistocene. *Geographical Review of Japan, Ser. A,* 65, 180-194. (in Japanese)

Yamakawa, K. (1981): Block streams and valley fill deposits in Kobugahara, Ashio maountains. *Memoirs of Geography, Graduate School, Hosei University,* 7, 51-64. (in Japanese).

Yanagimachi, O. (1987): Last Glacial Paleoclimate Reconstracted from Glacial and Periglacial Landforms in Central Japan. *Quaternary Research,* 25, 295-303. (in Japanese)

Washburn, A. L. (1947): Reconnaissance geology of portions of Vivtoria Island and adjacent regions. *Arctic Canada. Mem. Geol. Soc. Am.*, 22.

Wilhelmy, H. (1958) : *Klimamorphologie der Massengesteine*. Westermann, Braunschweig.

Williams, R. G. B. (1968): Some estimates of periglacial erosion in southern and eastern England. *Biuletyn Periglacjalny*, 17, 311-335.

In: Weathering: Types, Processes and Effects ISBN 978-1-61324-280-3
Editor: Matthew J. Colon, pp. 129-165

Chapter 3

WEATHERING PROCESSES AND NATURAL RADIONUCLIDES

Daniel Marcos Bonotto* and Jairo Roberto Jiménez-Rueda

Departamento de Petrologia e Metalogenia, Universidade Estadual Paulista (UNESP), Câmpus de Rio Claro, Av. 24-A No.1515, C.P. 178, CEP 13506-900, Rio Claro, São Paulo, Brasil

ABSTRACT

In managing our hydrological systems, increasing use is being made of information generated from studies held for understanding the controls on chemical weathering due to its importance in affecting the evolution of the Earth's surface, shaping landscapes, determining nutrient supply to ecosystems, and regulating global chemical cycles. A lot of recent studies have been realized to determine the parameters controlling the denudation, chemical weathering and physical erosion under different climatic conditions. In order to estimate chemical weathering fluxes affecting the drainage basins, models utilizing the Na, Ca, K, Mg and total dissolved load concentrations have been applied to obtain fractions coming from dissolved rocks, after corrections for river waters of the atmospheric inputs, dissolved solids from ion exchange sites in clays minerals, dissolved solids due to changes in biomass and

* E-mail: danielbonotto@yahoo.com.br

anthropogenic inputs, among other factors. Since uranium is among the main elements contributing to natural terrestrial radioactivity, several descendants of the mass number (4n+2) ^{238}U decay series have also been utilized in studies focusing weathered soil covers and hydrological resources in drainage basins. The ^{238}U decay series finishes at the stable ^{206}Pb and the radionuclides ^{238}U and ^{234}U have been used to evaluate the chemical weathering rate in soil profiles and hydrographic basins based on the fact that ^{234}U is preferentially mobilized to ^{238}U when rock weathers in the alteration front. Measurements of the $^{234}U/^{238}U$ activity ratio (AR) in rocks, soils and waters have allowed the calculation of the solution coefficient for the uranium characteristic of the region, that allows evaluate the time necessary to weather 1 m of rock under actual climatic conditions. This chapter summarizes some studies realized in a soil profile occurring at the giant Paraná sedimentary basin, Brazil, which have taken into account the presence of natural radionuclides belonging to the ^{238}U decay series.

INTRODUCTION

Weathering is an important phenomenon of the geochemical cycle as it contributes to the relief formation. It corresponds to a general term applied to physical and chemical changes suffered by rocks as a consequence of their exposition to different conditions of humidity and temperature (Bertin, 1974). Weathering is also considered a process related to rock disintegration or decomposition that is a common phenomenon occurring in the Earth's surface where the dominant factor is the rock interacting with water (Netto, 1980).

Generally speaking, weathering is often differentiated in two types: 1) disintegration- process of mechanical weathering in which the rocks are gradually fragmented in smaller pieces; 2) decomposition- process of chemical weathering resulting of reactions (the most common are oxidation, hydration, carbonation and dissolution) among the atmosphere constituents and rocks. Thus, reduction of the rock in fragments of smaller sizes is generated due to the mechanical weathering, whereas the chemical weathering causes the production of new substances/compounds (Bertin, 1974). However, the two processes occur simultaneously in nature and it is difficult or impossible to separate them, despite in some areas sometimes it is verified the dominance of one of them.

In general, the weathering processes are slow and their type and duration are controlled by climate. Therefore, it is expected that the chemical weathering becomes accelerated under humid and hot climatic conditions, whereas the decomposition would take place very slowly in more arid climate, which would favor the mechanical effects. Different rock types are also expected to imply in variable weathering rates. For instance, the weathering should occur in a higher rate in areas dominated by more soluble rocks like limestone and marbles in relation to regions characterized by the dominance of crystalline rocks.

After being weathered and disaggregated, the rock matrix supply fragments that can originate soils or constitute sediments, i.e. deposits of solid materials formed by a mobile agent (wind, ice or water) in the Earth's surface (Laporte, 1969).

When the soil formation takes place, there is a combination of effects of physical, chemical, biological, and anthropogenic processes on soil parent material. Thus, the soil genesis involves processes that develop layers or horizons in the soil profile, which cause additions, losses, transformations and translocations of material that compose the soil. Additionally, minerals derived from weathered rocks undergo further changes that cause the formation of secondary minerals and other compounds that are variably soluble in water. Such constituents are often moved from one area of the soil to others by water and animal activity so that the alteration and movement of materials within the soil causes the formation of distinctive soil horizons.

The sediments are variable, for example, sand beaches, lacustrine muds, water course pebbles, coral reefs and desert dunes (Laporte, 1969). The sediments are formed inside or outside of the occurrence area where occurs their final deposition, being water the main agent for their transportation. During its movement, the water flow transports sedimentary materials by three major pathways: 1) the water *dissolves* several substances like calcium, iron and carbonate ions; 2) materials of fine grain size and rock fragments are transported in *suspension* in the water course; 3) the water flow transports particles of gross grain size by *traction*, i.e. through jumping and saltation along the channel of the water course or river (Laporte, 1969). The sediments transported in suspension or by traction occasionally deposit in a delta at the river mouth, but the majority of the dissolved load is discharged into the sea where can precipitate or remain into solution (Laporte, 1969).

These aspects indicate that the sediments also belong to the hydrological cycle due to the importance of water on their transportation from the continents up to the oceans. A great diversity of factors acting under different time scales conduct to a differential accumulation of sediments in the several environments. Therefore, the evaluation of weathering and sedimentation rates provide more realistic estimate of the duration of the acting processes, constituting a topic of great interest in Geochemistry.

UNALTERED MAGMATIC ROCKS

The study of fresh/unaltered rocks is an important aspect to be considered when numerical indices are utilized for evaluating the weathering processes. Paul Niggli (1888–1953) was a swiss crystallographer and petrologist at the University of Zurich that studied magmas and metamorphic rocks, and introduced the use of phase diagrams in petrology. He also made a classification of sedimentary rocks according to grain shape and, in 1928, he and his students attempted to calculate the bulk chemical composition of the Earth.

Niggli proposed the grouping of the oxide components of an igneous rock according to their chemical similarity and defined several indices from the chemical analysis of a rock with several purposes, for instance, rock characterization, rock classification and comparison of rocks of the same petrographic province in order to determine their parental relationship and relative evolution grade (Mitchell and Bergman, 1991).

The first step on the method for generating the Niggli parameters is to calculate the molecular number (MN) of each oxide from the equation MN = 1000 WP / MW, where WP is the weight percent of the oxide and MW is its corresponding molecular weight. Table 1 summarizes some approximate MW values of typical oxides reported from chemical analyses of rocks.

Table 1. Approximate Molecular Weight (MW) of major oxides occurring in rocks

Oxide	WM	Oxide	WM
SiO_2	60	MgO	40
TiO_2	80	CaO	56
Al_2O_3	102	Na_2O	62
Fe_2O_3	160	K_2O	94
FeO	72	P_2O_5	104
MnO	71		

After determining the MN of each oxide, the value obtained for Fe_2O_3 is converted to FeO and summed to that found for FeO and such calculation is realized on multiplying the MN of Fe_2O_3 by two. Then, four Niggli parameters are defined: 1) $al = \Sigma$ NM ($Al_2O_3 + Cr_2O_3$ + REEs), where REEs are the rare earth elements; 2) $fm = \Sigma$ NM (FeO + MnO + MgO + NiO); 3) $c = \Sigma$ NM (CaO + BaO + SrO); 4) $alk = \Sigma$ NM ($Na_2O + K_2O + Li_2O$). The oxide groupings are initially calculated according to cation percentage and afterwards the sum for the rock has to be normalized to 100, i.e. $al + fm + c + alk = 100$. Each group corresponds to an individual Niggli parameter and each of which can be utilized for evaluating a silica value from the equation si = MN SiO_2 / $al + fm + c + alk$. Additionally, two more values were defined: a) k = MN K_2O/alk and b) mg = MN MgO/fm.

Beyond these seven Niggli parameters, a quartz index (qz) is calculated, whose importance is derived from the fact that the minerals more enriched in silica in the rocks are the feldspars and pyroxenes. The following proportion is verified in alkaline feldspars ($KAlSi_3O_8$) and calcic feldspars ($CaAl_2Si_2O_8$): $alk{:}al{:}si$ = 1:1:6 and $c{:}al{:}si$ = 1:1:2. However, the expected theoretical proportion in pyroxenes ($CaMgSi_2O_6$) is $fm{:}si = 1$ and $c{:}si = 1$. Therefore, the following relation exists in a rock composed by these minerals:

$$si' = 6\,alk + 2\,(al - alk) + 1\,[c - (al - alk)] = fm \tag{1}$$

The difference ($al - alk$) corresponds to the fraction of c bounded to alumina in feldspars, if $alk < al$. On the other hand, the difference $c - (al - alk)$ corresponds to the fraction of c that is uniquely associated to si. Now, if $al > alk + c$, then, the al excess is supposed to be combined with the same amount of si to constitute sillimanite. Remembering that $al+fm+c+alk=100$ and substituting in equation 1, it is possible to write:

$$si' = 100 + 4\,alk \tag{2}$$

However, if $alk > al$, then, the alkalis excess will form aegirine that implies on the equation:

$$si' = 100 + 3\,al + 1\,alk \tag{3}$$

When the value determined for si in a rock is higher than that found for si' (equation 2 or 3), the rock will contain free quartz (supersaturated rock). If *si* is lower than *si'*, then, the rock will contain minerals depleted in silica like olivine, biotite and feldspathoids. Therefore, the difference $qz = si - si'$ is an important index to evaluate the mineralogy of a rock that only may be known through its chemical analysis. Equation 2 may be applied if $al < alk + c$, whereas equation 3 can be utilized when the condition $alk > al$ is obeyed. If neither equation 2 nor equation 3 are applicable, then, $si' = 100$.

Table 2. Chemical composition (in weight percent) of some different rock types. Data reported by UNESP (1980)

Oxide	Syenite	Granite	Basalt	Carbonatite	Average crustal[1]
SiO_2	58.58	71.30	50.59	9.59	55.2
TiO_2	0.84	0.31	1.05	0.65	1.6
Al_2O_3	16.64	14.32	16.29	2.90	15.3
Fe_2O_3	3.04	1.21	3.66	4.33	2.8
FeO	3.13	1.64	5.08	4.37	5.8
MnO	0.13	0.05	0.17	0.72	0.2
MgO	1.87	0.71	8.96	6.69	5.2
CaO	3.53	1.84	9.50	34.06	8.8
Na_2O	5.24	3.68	2.89	1.02	2.9
K_2O	4.95	4.07	1.07	1.47	1.9
P_2O_5	0.29	0.12	0.21	1.86	0.3

[1]Poldervaart (1955).

Table 2 reports the results of chemical analysis for some rock types. Granite is a common and widely occurring type of intrusive, felsic, igneous rock. Granites usually have a medium- to coarse-grained texture. Occasionally some individual crystals (phenocrystals) are larger than the groundmass, in which case the texture is known as porphyritic. Syenite is a coarse-grained intrusive igneous rock of the same general composition as granite but with the quartz either absent or present in relatively small amounts (<5%). The feldspar component of syenite is predominantly alkaline in character (usually orthoclase) and the plagioclase feldspars may be present in small quantities, less than 10%. Basalt is a common extrusive volcanic rock that is usually grey to black and fine-grained due to rapid cooling of lava at the surface of a planet. It may be porphyritic, containing larger crystals in a fine matrix, or vesicular, or frothy scoria; unweathered basalt is black or grey. Carbonatites are intrusive or extrusive igneous rocks defined by mineralogic composition consisting of greater than 50 percent carbonate minerals; they may be confused with marble and often require geochemical verification for proper identification.

Tables 3-7 give the Niggli parameters and related parameters for the rock specimens reported in Table 2. It is evident that $qz > 0$ for the granite and the crust, i.e. the widespread presence of granites and other acid rocks at the Earth's surface imply on causing a slight quartz supersaturation in the crust, when the mean global composition is taken into account. The highly negative *qz* value for the carbonatite is compatible with the enhanced presence of carbonate minerals in these rocks and the negative *qz* values for the syenite and basalt samples also agree with the mineralogical assemblage of these basic rocks. Thus, the Niggli parameters are useful to study some major features of unaltered magmatic rocks and others indices will be helpful to study weathering processes affecting them, as described in the next sections.

Table 3. Niggli parameters for the syenite sample whose chemical composition is given in Table 2

Oxide	MN	Parameter	Value
SiO_2	976.50	*al*	163.14
Al_2O_3	163.14	*fm*	130.05
Fe_2O_3	19.00	*c*	63.00
FeO	43.47	*alk*	137.18
MnO	1.83	*al+fm+c+alk*	493.37
MgO	46.75	*si*	197.92
CaO	63.04	*si'*	211.20
Na_2O	84.52	*qz*	-13.30
K_2O	52.66		

Table 4. Niggli parameters for the granite sample whose chemical composition is given in Table 2

Oxide	**MN**	**Parameter**	**Value**
SiO_2	1188.33	*al*	140.40
Al_2O_3	140.39	*fm*	56.35
Fe_2O_3	7.56	*c*	32.86
FeO	22.78	*alk*	102.65
MnO	0.70	*al+fm+c+alk*	332.26
MgO	17.75	*si*	357.65
CaO	32.86	*si'*	100.00
Na_2O	59.35	*qz*	257.65
K_2O	43.30		

Table 5. Niggli parameters for the basalt sample whose chemical composition is given in Table 2

Oxide	MN	Parameter	Value
SiO_2	843.17	*al*	159.70
Al_2O_3	159.70	*fm*	342.68
Fe_2O_3	22.88	*c*	169.64
FeO	70.56	*alk*	57.99
MnO	2.39	*al+fm+c+alk*	729.89
MgO	224.00	*si*	115.52
CaO	169.64	*si'*	131.76
Na_2O	46.61	*qz*	-16.20
K_2O	11.38		

Table 6. Niggli parameters for the carbonatite sample whose chemical composition is given in Table 2

Oxide	MN	Parameter	Value
SiO_2	159.83	*al*	28.43
Al_2O_3	28.43	*fm*	292.21
Fe_2O_3	27.06	*c*	608.21
FeO	60.69	*alk*	32.09
MnO	10.14	*al+fm+c+alk*	960.74
MgO	167.25	*si*	16.60
CaO	608.21	*si'*	113.20
Na_2O	16.45	*qz*	-96.60
K_2O	15.64		

Table 7. Niggli parameters for the average crustal composition given in Table 2

Oxide	MN	Parameter	Value
SiO_2	920.00	*al*	150.00
Al_2O_3	150.00	*fm*	248.37
Fe_2O_3	17.50	*c*	157.14
FeO	80.56	*alk*	66.97
MnO	2.82	*al+fm+c+alk*	622.48
MgO	130.00	*si*	147.80
CaO	157.14	*si'*	143.20
Na_2O	46.77	*qz*	4.60
K_2O	20.20		

GRAIN SIZE AND WEATHERING SEQUENCES

When a mixture of minerals deposited together in one rock formation is subjected to the agencies of chemical weathering, some minerals are weathered faster than others. A "weathering sequence" corresponds to a list of minerals according to their relative stability to weathering and the fundamental basis of its establishment lies with the following criteria (Jackson and Sherman, 1953): relative persistence with age of formation, geographic correlation with weathering intensity factors, particle-size (specific surface) function, and persistence as a function of depth in the formation.

The grain size and its distribution is much important to express the weathering products and a terminology commonly utilized in the literature corresponds to that of the Wentworth (1922) scale, which includes the major classes: gravel, sand and clay, with their numerous subdivisions. Because the range of grain sizes found in nature is so large, a logarithmic scale, such as the Udden-Wentworth, is more practical than a linear scale, despite it inadequately covers gravel (boulder, cobble, pebble, and granule) that is the dominant fraction in many environments. The diameters of the particles are used in the Udden-Wentworth scale (often called the Wentworth scale) as a grade scale. Particles larger than 64 mm in diameter are classified as cobbles. Smaller particles are pebbles, granules, sand and silt. Those smaller than 0.0039 mm are clay. Table 8 shows the classes according to the Wentworth (1922) scale.

Table 8. Grain size scales

Size (mm)	**Size (μm)**	**Wentworth (1922) size class**
4096 - 256		Boulder
256 - 64		Cobble
64 - 4		Pebble
4 - 2		Granule
2 - 1	2000 - 1000	Very coarse sand
1 – 1/2	1000 - 500	Coarse sand
1/2 – 1/4	500 - 250	Medium sand
1/4 - 1/8	250 - 125	Fine sand
1/8 – 1/16	125 - 63	Very fine sand
1/16 – 1/32	63 - 31	Coarse silt
1/32 – 1/64	31 – 15.6	Medium silt
1/64 – 1/128	15.6 – 7.8	Fine silt
1/128 – 1/256	7.8 – 3.9	Very fine silt
< 1/256	< 3.9	Clay

For many authors investigating weathering processes in soil profiles, the term "clay fraction" includes the particles less than 2 μm in equivalent diameter, whereas the term "mineral colloid" refers to the finer clay fractions, including particles less than 0.2, 0.1, 0.08, etc. μm in diameter (Jackson and Sherman, 1953).

Different weathering sequences of minerals have been worked out for different categories of minerals, i.e. those exhibiting heavy specific gravity, coarse-grained minerals, colloidal minerals, and combinations of these categories (Jackson and Sherman, 1953). Pettijohn (1941) found that the less stable minerals of heavy specific gravity occur in decreasing amounts in rocks of increasing age and his weathering sequence (most stable mineral first) is: anatase, muscovite, rutile, zircon, tourmaline, monazite, garnet, biotite, apatite, ilmenite, magnetite, staurolite, kyanite, epidote, hornblende, andalusite, topaz, spene, zoisite, augite, sillimanite, hypersthene, diopside, actinolite and olivine. Weyl (1952) grouped the heavy minerals in four categories of stability: a) extremely unstable- olivine, hornblende, augite; b) slightly stable- garnet, epidote; c) stable- staurolite, kyanite, sillimanite, andalusite; d) very stable- tourmaline, zircon, rutile, titanite, magnetite. The Pettijohn's sequence indicates that zircon is a most appropriate mineral rather than alumina to calculate the chemical weathering losses of materials. However, the zirconium analyses are not much frequent like the alumina ones, reason by which it is commonly employed.

Goldich (1938) presented a branched stability series of coarse-grained minerals (most stable last) as reported in Table 9. This sequence is similar to the Bowen (1922) sequence for temperature of crystallization, i.e. olivine at 1890 °C grading to quartz at 570 °C, a fact that has been interpreted as reflecting the increasing stability of minerals with the increasing departure of the temperature from that of the environment of formation. Quartz is listed as the most stable mineral in the sequence because coarse-grained quartz is extremely resistant to weathering. Humbert and Marshal (1943) have shown how quartz grains can be weathered by iron-oxide crystal growth in cracks. Engelhardt (1937) similarly noted sharp-angled (fractured) quartz grains in the finer fraction and rounded grains in the coarse. Quartz particles of clay size have a lower relative stability.

Table 9. The stability series of coarse-grained minerals, according to Goldich (1938)

Mafic	Feldspathic
Olivine	Calcic plagioclase
Augite	Calci-alkalic plagioclase
Hornblende	Alkali-calcic plagioclase
Biotite	Alkalic plagioclase
Potassium feldspar	
Muscovite	
Quartz	

With increasing fineness of mineral particles, the mineral stability sequence is somewhat different from that of coarse-grained mineral particles because specific surface becomes great enough to hasten weathering of minerals which are more stable in coarser sizes (Jackson and Sherman, 1953). Jackson and Sherman (1953) pointed out the following mineral weathering sequence (most stable stage given last) for fine particles (less than 2 μm in diameter and in the size range of 2-5 μm corresponding to fine silt-size particles): 1) gypsum (also halite, sodium nitrate and ammonium chloride); 2) calcite (also dolomite, aragonite and apatite); 3) olivine-hornblende (also pyroxenes and diopside); 4) biotite (also glauconite, magnesium chlorite, antigorite and nontronite); 5) albite (also anorthite, stilbite, microcline and orthoclase); 6) quartz (also cristobalite); 7) muscovite (also sericite and illite); 8) interstratified 2:1 layer silicates and vermiculite (including partially expanded hydrous micas, randomly interstratified 2:1 layer silicates with no basal spacings and regularly interstratified 2:1 layer silicates); 9) montmorillonite (also beidellite and saponite); 10) kaolinite (also halloysite); 11) gibbsite (also boehmite and allophane); 12) hematite (also goethite and limonite); 13) anatase (also zircon, rutile, ilmenite, leucoxene and corundum).

Some Weathering Indices

Several indexes have been reported in the literature for expressing the degree of chemical weathering of a mineral or soil material and some of them will be considered in this chapter. Ratios of moles of different constituents of rocks, minerals and soils have been employed to condense elemental analyses of these materials, for instance (Jackson and Sherman, 1953): silica:alumina ratio (SiO_2/Al_2O_3), silica:ferric oxide ratio (SiO_2/Fe_2O_3), silica:sesquioxide ratio ($SiO_2/Al_2O_3+Fe_2O_3$), bases:alumina ratio (K_2O+Na_2O+CaO/Al_2O_3), etc. Alumina has often been employed as the basis for calculation of relative weathering losses of other constituents and, hence, the referred ratios serve as measures of depletion of the other constituents. In many soil profiles, ratios involving alumina are calculated considering the leached horizon and the parent material.

Table 10 reports the results of the chemical analyses in soil samples collected in 6 points located at the rural zone of Patrocínio Paulista, São Paulo State, Brazil. LOI (Loss on Ignition) expresses the organic matter + adsorbed water + water in crystal lattices and fluid inclusions + CO_2 of carbonates + SO_2 of sulfides. The soils analyzed indicate that has occurred polycyclic slow mass flow (gravitational) in the region, with accumulation of detritic concretions (lateritic/pedorelict). Such detritic flow is possibly related to the erosion of latosolized/lateralized landscapes from Lower Tertiary to Intermediary Pleistocene, which has been reworked during their evolution, forming neosols, argisols and cambisols. They can't be considered latosols and exhibit signs of previous laterization processes despite the dominance of the tropical climatic conditions in more recent times have favored the occurrence of ferrolysis.

In lateritic soils formed by intense weathering conditions, the Si concentration may be diminished, enhancing the Al-Fe content. In general, the evolution of laterisation processes is related to the nature of the original rock matrix and involves all Al-silicates rock types. The chemical composition of the soils given in Table 10 reflects at least some of these aspects, as indicated the inverse relationships among SiO_2, Al_2O_3, Fe_2O_3, TiO_2, MgO and P_2O_5 (Figure 1).

Table 10. Results of the chemical analyses in soil samples collected at Patrocínio Paulista, São Paulo State, Brazil

Sample Code	Depth (cm)	SiO_2	TiO_2	AL_2O_3	Fe_2O_3	MnO	MgO (wt. %)	CaO	Na_2O	$P2O_5$	LOI
A1	0-20	12.31	9.56	28.64	29.38	0.11	0.37	0.13	0.02	0.66	18.77
A1	20-60	11.97	9.74	28.86	29.92	0.11	0.37	0.11	0.02	0.61	18.22
A2	0-20	14.24	9.14	28.62	29.09	0.12	0.30	0.23	0.02	0.56	17.63
A2	20-60	14.16	9.39	29.58	30.04	0.12	0.30	0.17	0.03	0.49	15.69
A3	0-20	19.73	8.65	27.73	37.43	0.11	0.28	0.15	0.02	0.42	15.44
A3	20-60	19.36	8.49	27.65	27.15	0.09	0.26	0.07	0.02	0.29	16.58
A4	0-20	23.39	7.88	25.54	24.62	0.09	0.26	0.22	0.02	0.55	17.37
A4	20-60	22.27	8.21	26.40	25.58	0.09	0.27	0.12	0.03	0.40	16.57
A5	0-20	31.60	7.30	22.46	24.55	0.12	0.19	0.03	0.03	0.19	13.49
A5	20-60	29.93	7.65	23.64	24.98	0.10	0.20	0.03	0.02	0.19	13.24
A6	0-20	15.33	9.00	27.90	27.39	0.09	0.28	0.19	0.03	0.48	19.26
A6	20-60	15.79	9.14	27.88	28.54	0.10	0.25	0.07	0.02	0.38	17.79

According to Waal et al. (2009).

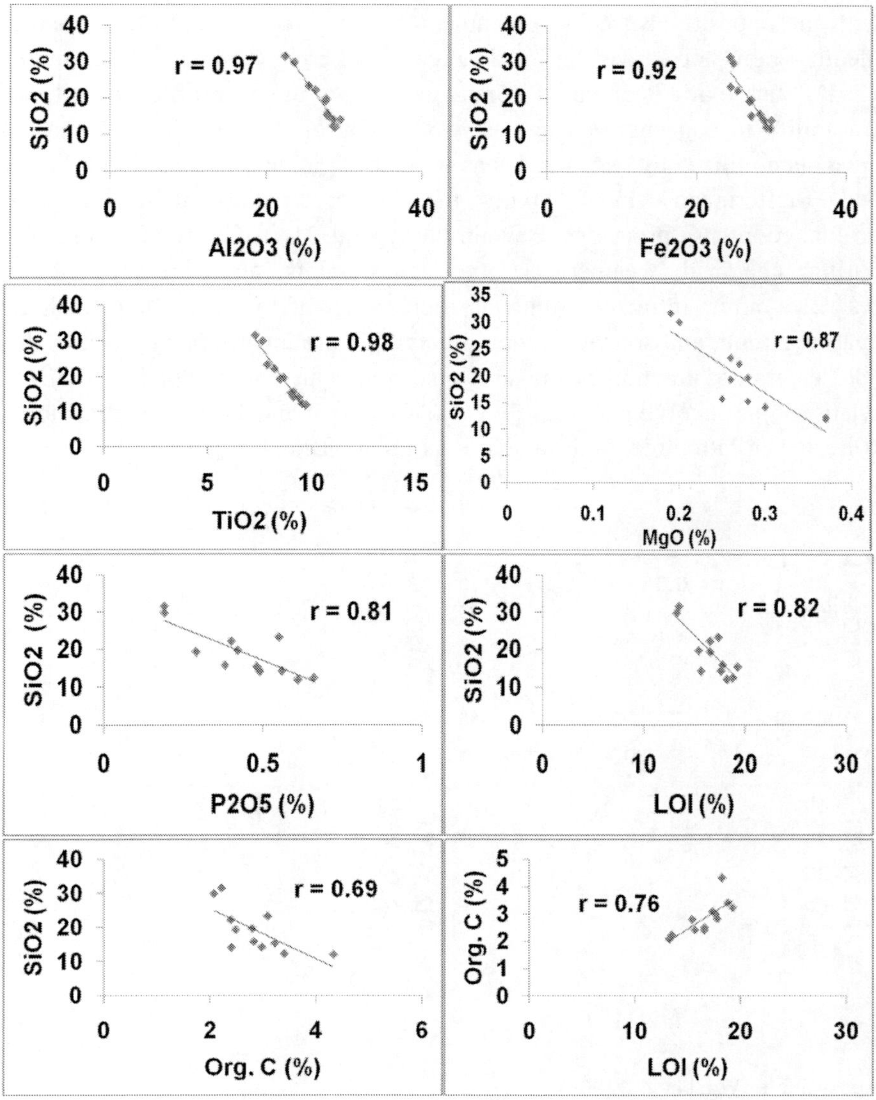

Data on organic C are reported by Waal et al. (2009).

Figure 1. Relationships among SiO_2, Al_2O_3, Fe_2O_3, TiO_2, MgO, P_2O_5, LOI and organic C in soil samples collected at Patrocínio Paulista, São Paulo State, Brazil.

Thus, Si-poorer sediments are more Al-Fe-Ti-Mg-P-enhanced, suggesting the occurrence of more prolonged/intense weathering processes, with the opposite also being probable. Al_2O_3 is a constituent often used to identify gains and losses during chemical weathering (Colman, 1982; Faure, 1991). Such oxide is assumed to remain constant in soil profiles, despite its variability in concentration against depth (Faure, 1992). SiO_2/Al_2O_3 ratios have been utilized by geochemists as weathering indices (Pedro, 1968; Pedro and Sieffermann, 1979), with the highest values indicating less prolonged/intense processes. Beyond the typical Al_2O_3-Fe_2O_3 accumulations during chemical weathering, other constituents also are present in resistates/minor refractory minerals more resistant to weathering (such as apatite, zircon, and sphene, among others) that are incorporated intact in the coarser grained fraction of the soils analyzed. This could explain the direct relationships involving Al_2O_3, Fe_2O_3, TiO_2, MgO and P_2O_5 in soil samples collected at Patrocínio Paulista, São Paulo State, Brazil (Figure 2).

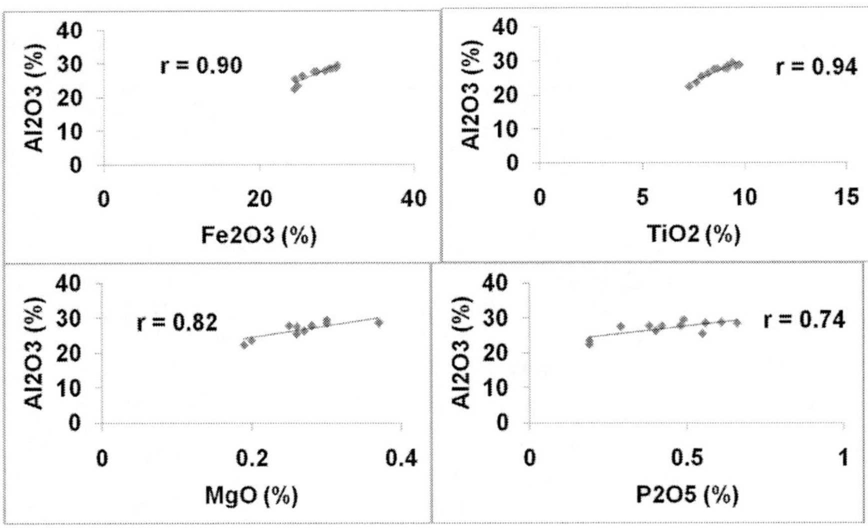

According to Waal et al. (2009).

Figure 2. Relationships between Al_2O_3 and Fe_2O_3, TiO_2, MgO and P_2O_5 in soil samples collected at Patrocínio Paulista, São Paulo State, Brazil.

The direct relationship between LOI and organic C (Figure 1) is related to the fact that LOI is a parameter linked to the organic matter (organic C) content of the soils. Beyond modifying the bulk density and porosity of soils, organic matter also influences their specific surface, since 1% of organic matter in porous matrices may cause an increase of the specific surface in about 7 m^2 (Kiehl, 1977). Such increase may raise the cations exchange capacity, cations adsorption, and percentage of water retention in soils. Silica is inversely related to organic matter (and organic C) at the area studied (Figure 1), being expected that the decrease in SiO_2 concentration implies on an increase of the soils specific surface.

Figure 3. Location of the Paraná sedimentary basin in South America.

WEATHERING OF BASALT OF THE PARANA SEDIMENTARY BASIN

The Paraná sedimentary basin is a giant geological unit whose surface area corresponds to 1.6×10^6 km^2 and comprises southern Brazil (1×10^6 km^2 in the states of Mato Grosso, Mato Grosso do Sul, Goiás, Minas Gerais, São Paulo, Paraná, Santa Catarina and Rio Grande do Sul), eastern Paraguay (0.1×10^6 km^2), NW Uruguay (0.1×10^6 km^2) and the northeastern extreme corner of Argentina (0.4×10^6 km^2) (Figure 3) (Araújo et al., 1999). The basin is intercratonic, where the sedimentary sequence covering since the Silurian-Devonian up to the Cretaceous periods is almost undisturbed, with gentle dips towards the center of the basin.

One very important lithological unit in Paraná sedimentary basin is the Serra Geral Formation that constitutes a designation proposed by White (1908) to the group of basalts formed by flows through extensive fissure volcanism, estimated in 650,000 km^3, that are associated to intrusive bodies of the same composition, mainly constituting dykes and sills (Almeida and Melo, 1981). There is an area exhibiting several diabase dykes at northeast of São Paulo State, close to the border of the basin, where also occurs many sills. The flows are formed by dark gray to black colored afanitic rocks, with variable thickness (up to 50-100 meters) and horizontal extension exceeding 10 km. The central zone of the thicker flows is massif, microcrystalline, fractured by sub-vertical contraction joints that divide rock in columns. The upper portion of the flows contains vesicles and amigdales that are often horizontally elongated and contain the dominant percentage of the vitreous matter in the rock. The amigdales are partially or completely filled by chalcedony, quartz, calcite, zeolite and nontronite that is a mineral able to supply a greenish color. The basal zone of the flows exhibits similar aspects despite the abundance and thickness are slightly more reduced. Horizontal joints are frequent in the basal portions and top of the large flows that represent the laminar lava flow in the interior of the spills (Bagolini, 1971). In terms of petrographic features, the basalts of the Serra Geral Formation are primarily constituted by zoned labradorite associated to clinopyroxenes (augite and sometimes pigeonite), and secondarily by titanomagnetite, apatite, quartz and rare olivine and respective alteration products. The absence of Botucatu sandstone in the Serra Geral Formation has been suggested due to the difficulty of performing distinction between the flows and sills. However, Almeida and Melo (1981) pointed out several evidences of contemporary sedimentation and volcanism, indicating that the Botucatu Formation represents the various sub-environments of a large climatic desert of increasing dryness, whose existence prolonged up to the occurrence of the basaltic volcanism. The main phase of the magmatic manifestation occurred between 130 and 115 million years ago, as indicated the results obtained for samples dated by the potassium-argon method in whole rock (Amaral et al., 1966; Melfi, 1967). Additional estimate of the volcanism age by the ^{40}Ar-^{39}Ar method was performed by Renne et al. (1992) and Turner et al. (1994) that provided values between 137 and 127 Ma, which do not differ significantly of those reported by Amaral et al. (1966) and Melfi (1967). Turner et al. (1994) also suggested that the duration of the volcanism was 10 million years and estimated a low lava effusion rate corresponding to 0.1 km^3/year.

Faure (1991) described the effects of weathering on the chemical composition of a basalt from the state of Paraná, Brazil, and for a sample of saprolite that formed from it, whose data are given in Table 11. The saprolite has lower concentrations of SiO_2, CaO, MgO, Na_2O, K_2O, MnO and P_2O_5 than the basalt. However, it has higher concentrations of Al_2O_3, Fe_2O_3, TiO_2, and LOI. Faure (1991) chose Al_2O_3 as the constant oxide despite its concentration apparently increased from 12.3% tom 26.3% as a result of weathering of the basalt. Such constraint implied on the adoption of a weight-loss factor (w) derived from the ratio of the concentrations of the constant oxide in the fresh and weathered rock, i.e. $w = (Al_2O_3)_{fresh}/(Al_2O_3)_{weath.}$. According to data reported in Table 11, $w = 12.3/26.3 = 0.467$, and this factor allows estimate the amount of all oxides remaining after weathering (see column 4 in Table 11). The actual gains and losses of each constituent can be determined by comparing the amount remaining to the analysis of the basalt and such values can be expressed in terms of percent of the amount originally present in the basalt, as also reported in Table 11. Thus, all parameters obtained involving the basalt and saprolite permit to point out that for 100 g of rock:

- SiO_2 is lost (34.3 g), as well Fe_2O_3 (5.06 g) and TiO_2 (1.34 g);
- Fe_2O_3 is lost despite its concentration in the saprolite is higher (21.5%) than in the basalt (15.1%);
- TiO_2 is lost despite its concentration in the saprolite is higher (3.96%) than in the basalt (3.19%);

- All constituent are lost, except LOI (H_2O, CO_2 and SO_2), relative to constant Al_2O_3;
- The greatest losses occurred for CaO and Na_2O (99.6%), MgO (97.1%), K_2O (96.5%), and P_2O_5 (89.1%) followed by MnO (75.0%), SiO_2 (67.6%), TiO_2 (42.0%), and Fe_2O_3 (33.5%);
- The basalt weathering led to the formation of saprolite primarily composed of Al_2O_3, SiO_2, Fe_2O_3, and H_2O.

Table 11. Comparison of chemical analyses of basalt from Paraná State, Brazil, and its weathering products

Oxide	Basalt (%)	Saprolite (%)	Amount remaining (g)	Gain (+) or Loss (-) (g)	Gain (+) or Loss (-) (%)
SiO_2	50.70	35.10	16.4	-34.30	-67.6
Al_2O_3	12.30	26.30	12.3	0	0
CaO	7.83	0.06	0.028	-7.80	-99.6
MgO	4.18	0.26	0.12	-4.06	-97.1
Na_2O	2.53	0.02	0.009	-2.52	-99.6
K_2O	1.71	0.13	0.061	-1.65	-96.5
Fe_2O_3	15.10	21.50	10.04	-5.06	-33.5
MnO	0.20	0.10	0.046	-0.15	-75.0
TiO_2	3.19	3.96	1.85	-1.34	-42.0
P_2O_5	0.46	0.11	0.051	-0.41	-89.1
LOI	1.70	12.40	5.79	+4.09	+241
Sum	100.00	100.00	46.7	-53.30	

According to Faure (1991).

WEATHERING OF SANDSTONE OF THE PARANA SEDIMENTARY BASIN

Like other sedimentary sequences, the Paraná basin has also been extensively investigated due to its economic potential for supplying water, petroleum, coal, clays, and metals like uranium, among other resources. The Tubarão Group in Paraná basin contains records of the Permo-Carboniferous glaciation, marine ingressions, flora relicts and coal layers. The paleoenvironments related to these deposits have suffered much vertical and horizontal variation and, therefore, the sub-division of the Group is much difficult when applied to large areas. Almeida and Melo (1981) considered that in São Paulo State there are no enough elements for justifying the sub-division of the Tubarão Group beyond of the following Formations: Itararé, Aquidauana and Tatuí. The sediments of the Itararé Formation (Subgroup) primarily constitute fine to coarse sandstones, mudstones and diamictites whose colors are yellow, red and clear gray in the upper and lower parts; the intermediary portion is represented by fine sandstones, siltstones and mudstones whose colors are dark gray and yellow. The sandstones are frequently feldspathic or ***arcosian*** and form psamitic bodies. They exhibit syngenetic sedimentary structures like ***wave***-formed ***ripple*** marks, crossed and gradational stratification, and also structures due to plastic deformation paene contemporary to deposition (DAEE, 1981). The Aquidauna Formation is mainly constituted by sandstones and siltstones of reddish color despite also occurs red or greenish shales, conglomerates, diamictites and rhythmites. The fine sandstones and siltstones constitute tabular or elongated bodies with intercalations of lenses of coarse feldspathic sandstone, pebbles, conglomerate lenses with a sandy matrix and centimetric pebbles of rocks of the crystalline basement. The sandstones exhibit plane-parallel, layered, tangential or crossed stratification when are not massif. The Tatuí Formation is mainly constituted by siltstones despite also occurs layers of sandstones, limestones, shales and sílex. These rock types represent uniform sedimentation that contrasts with the typical heterogeneity of the Itararé Subgroup (Almeida and Melo, 1981).

It has been investigated a soil profile developed over the Itararé Subgroup and located in the 50-km wide Depressão Periférica geomorphological province at São Paulo State, Brazil. The studied area is situated in Limeira city, about 150 km distant from São Paulo city (Figure 4). Several detritic lithofacies comprise the Itararé Subgroup, where the main lithological types are massive feldspathic or arkoseous sandstones of fluvial, marine, lacustrine, deltaic or eolic origin, varying considerably in texture and fossil content. It was sampled the weathered cover developed over an altered sandstone matrix 17-m deep and 6 km distant from Limeira city.

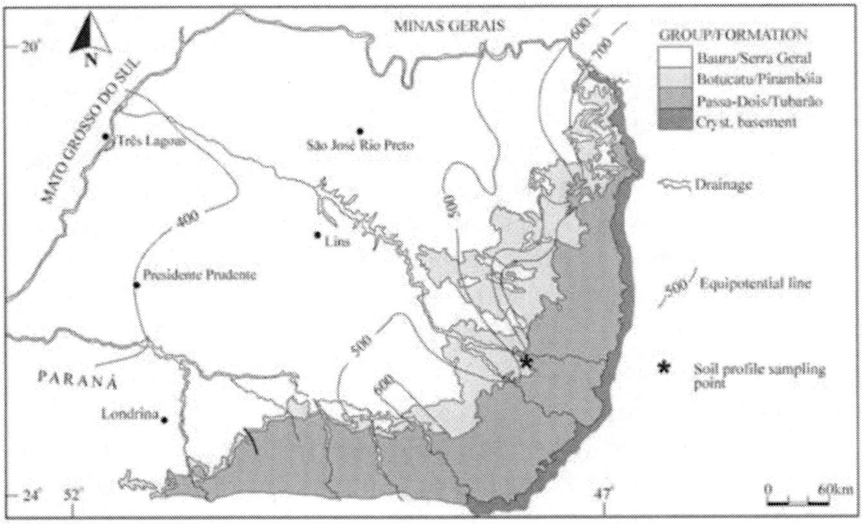

Figure 4. Location of the soil profile sampled at Itararé Subgroup, São Paulo State, Brazil.

Some 2 kg of residual rock and soil samples representative of different horizons in the profile developed over Itararé Subgroup were collected with a 30-cm long, 10-cm diameter galvanized steel auger bucket connected to a cross handle by a 1-m long extension. The samples were described in the field, placed in plastic bags, sealed, and after drying and mixing, aliquots were separated for mineralogical, physical and chemical identification.

X-ray diffraction was utilized in the mineralogical identification of the samples, whereas the major element analysis was performed by X-ray fluorescence spectrography, and the organic matter by colorimetry (Allman and Lawrence, 1981). The description of the samples and their physical and chemical characterization is shown in Figure 5 and Table 12. The presence of plintite (Federici and Pappalardo, 2001) enriched of iron oxides was identified to occur mixed with the residual rock; it was supplied by oxidation and reduction alternately occurring as a consequence of periodical oscillation of the water table over the years, thus indicating long exposition to a warm and damp climate, typically tropical, for a time span long enough to permit the soil evolution (Federici and Pappalardo, 2001).

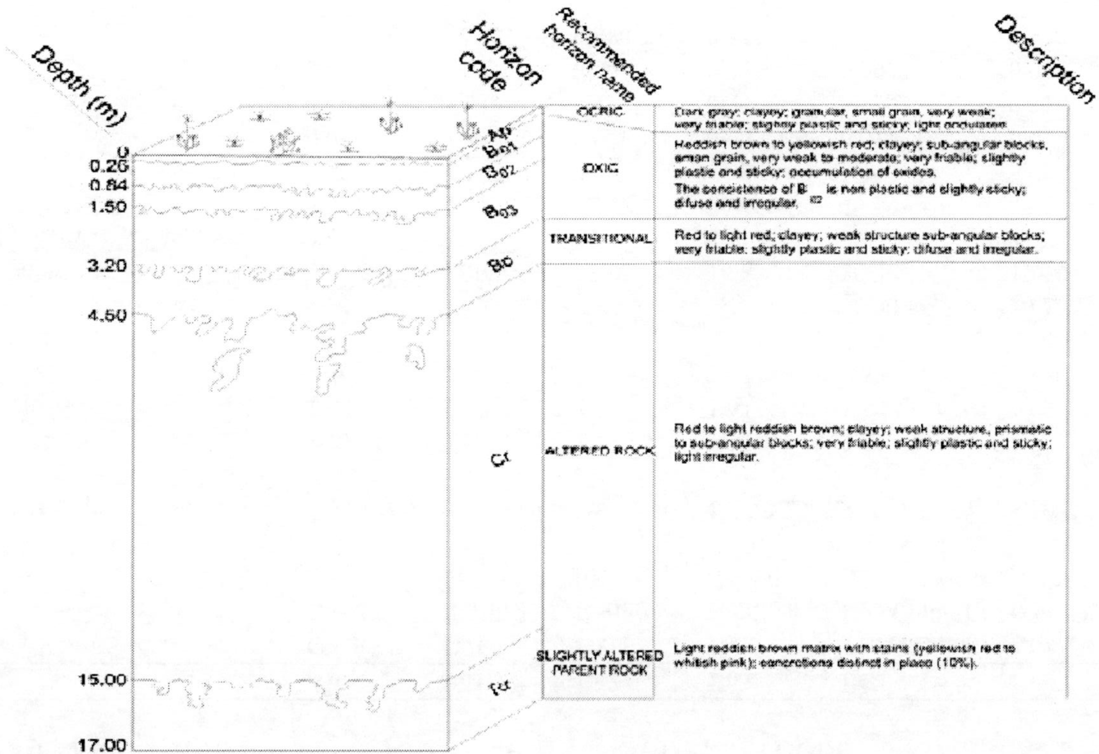

Figure 5. Description of the soil profile developed over sandstone from Itararé Subgroup, which was sampled at Limeira city, São Paulo State, Brazil.

Table 12. Physical and chemical evaluation of the soil profile developed over sandstone from Itararé Subgroup, São Paulo State, Brazil

PARAMETER	UNIT	HORIZON CODE						Residual Rock
		Ap	Bo1	Bo2	Bo3	BC	Cr	
Horizon Name	-	**Ocric**	**Oxic**	**Oxic**	**Oxic**	**Transitional**	**Altered Rock**	
Color	-	**Dark gray**	**Reddish brown**	**Reddish brown**	**Reddish brown**	**Red to light red**	**Red to light reddish brown**	**Light reddish brown**
Interval of Depth	m	0.00-0.26	0.26-0.84	0.84-1.50	1.50-3.20	3.20-4.50	4.50-7.00	15.00-17.00
Average Depth	m	0.13	0.55	1.17	2.35	3.85	5.75	16.00
Bulk Density	g/cm^3	1.80	1.73	1.50	1.61	1.58	1.65	2.11
True Density	g/cm^3	2.56	2.58	2.54	2.63	2.60	2.57	2.50
Total Sand	wt. %	86.82	87.83	86.07	83.46	79.78	80.40	-
Total Silt	wt. %	7.31	6.25	7.84	9.18	10.00	10.20	-
Total Clay	wt. %	13.17	12.67	13.92	16.53	20.22	19.59	-
Total Porosity	wt. %	29	33	41	39	39	36	16
SiO_2	wt. %	67.77	81.64	72.68	80.65	79.59	85.27	82.23
Al_2O_3	wt. %	3.83	4.96	4.94	7.38	8.39	5.78	2.82
Na_2O	wt. %	0.016	0.057	0.040	0.045	0.076	0.069	0.073
K_2O	wt. %	0.10	0.10	0.12	0.21	0.33	0.31	0.44
CaO	wt %	0.056	0.053	0.048	0.066	0.092	0.052	0.068
MgO	wt. %	0.065	0.092	0.084	0.139	0.191	0.127	0.054

PARAMETER	UNIT	HORIZON CODE						Residual Rock
		Ap	Bo1	Bo2	Bo3	BC	Cr	
Horizon Name	-	**Ocric**	**Oxic**	**Oxic**	**Oxic**	**Transitional**	**Altered Rock**	
Color	-	**Dark gray**	**Reddish brown**	**Reddish brown**	**Reddish brown**	**Red to light red**	**Red to light reddish brown**	**Light reddish brown**
Fe_2O_3	wt. %	1.40	1.60	1.65	2.15	2.54	1.22	0.44
MnO	wt. %	0.007	0.008	0.006	0.008	0.012	0.008	0.004
TiO_2	wt. %	0.35	0.40	0.44	0.53	0.58	0.36	0.05
P_2O_5	wt. %	0.013	0.018	0.013	0.020	0.022	0.016	0.007
Organic Matter	wt. %	1.29	0.82	0.62	0.47	0.34	0.28	0.37
Other[1]	wt.%	26.39	11.07	19.98	8.79	8.19	6.78	13.81

[1] Other elements and compounds not analyzed, with predominance of adsorbed water, water in crystal lattices and fluid inclusions, CO_2 of carbonates, and SO_2 of sulfides.

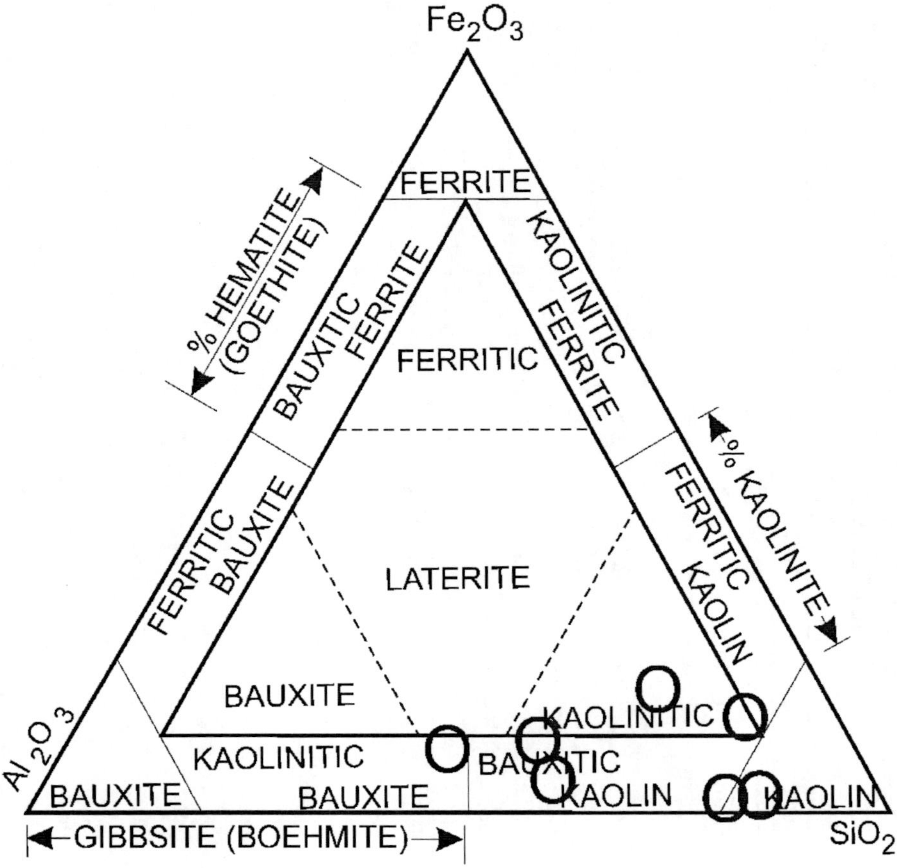

Figure 6. Data for the soil profile developed over sandstone from Itararé Subgroup, São Paulo State, Brazil, plotted on the SiO_2-Al_2O_3-Fe_2O_3 diagram proposed by Balasubramaniam et al. (1983).

The analysis of the particle size distribution indicated the predominance of sand in all samples (80-88 wt.%), and an excellent inverse correlation ($r = -1$) between the sandy and clayey fractions. The insertion of the chemical data obtained for the analyzed samples in the SiO_2-Al_2O_3-Fe_2O_3 triangle proposed by Schellmann (1979) for evaluating the degree of weathering indicated that all horizons fall into the field of the sandstone matrix dominantly composed by SiO_2. Furthermore, the insertion of the same data in the diagram proposed by Balasubramaniam et al. (1983) for defining the presence of ferrite, bauxite, and kaolin (Figure 6) shows that kaolinization processes are taking place in the soil horizons and altered parent rock. Such finding was confirmed by the use of X-ray diffraction in the mineralogical identification of the samples, which indicated the presence of kaolinite in all them.

Because Al_2O_3 is the constituent most often chosen for identifying real gains and losses during chemical weathering (Colman, 1982; Faure, 1991), this oxide was considered to remain constant in amount in the soil profile described in Table 12, despite its concentration appeared to have changed. Under this assumption, it has been possible to compare the chemical analyses of the residual rock from Itararé Subgroup and its weathering products in Cr and BC horizons, similarly to the comparison realized from data reported in Table 11. The results of the calculation are shown in Table 13, which allow verify that Mg, Fe, and Ti have accumulated in Cr and BC horizons relatively to the residual rock.

Significant correlations between Al_2O_3 and MgO ($r = 0.97$), Al_2O_3 and P_2O_5 ($r = 0.92$), Al_2O_3 and Fe_2O_3 ($r = 0.90$), Al_2O_3 and MnO ($r = 0.88$), and Al_2O_3 and TiO_2 ($r = 0.87$) (Figure 7) have been found in all soil horizons. Micas, feldspars, clays, apatite, and amorphous Fe-Mn oxi-hydroxides are the main probable sources of these constituents. Profuse rainfall (about 1.6cm/year) characterizes the studied area, favoring the occurrence of substantial losses of silica and bases in all soil horizons relatively to the slightly altered parent rock, i.e. 39-67 wt.% SiO_2, 55-84 wt.% Na_2O, 66-87 wt.% K_2O, and 39-63 wt.% CaO. The remaining Na_2O correlated significantly with SiO_2 ($r = 0.85$) and K_2O ($r = 0.78$), possibly reflecting the presence of feldspars, micas and clays as their source minerals.

U-Isotopes and Weathering Processes

Cosmic-ray produced radionuclides (like ^{7}Be, ^{10}Be and ^{26}Al) coupled to geochemical mass balance methods have been recently used for investigating long-term denudation/erosion/weathering processes affecting hydrographic basins (for instance, Baskaran, 1995; Small et al., 1999; Riebe et al., 2000, 2003, 2004).

Since uranium is among the main elements contributing to natural terrestrial radioactivity, several descendants of the mass number (4n+2) ^{238}U decay series have also been utilized in studies focusing weathering processes. The ^{238}U decay series finishes at the stable ^{206}Pb, according to the sequence (Figure 8): ^{238}U (4.49 Ga, α) $\rightarrow$ ^{234}Th (24.1 d, β^-) $\rightarrow$ ^{234}Pa (1.18 min, β^-) $\rightarrow$

^{234}U (0.248 Ma, α) → ^{230}Th (75.2 ka, α) → ^{226}Ra (1622 a, α) → ^{222}Rn (3.83 d, α) → ^{218}Po (3.05 min, α) → ^{214}Pb (26.8 min, β^-) → ^{214}Bi (19.7 min, β^-) → ^{214}Po (0.16 ms, α) → ^{210}Pb (22.26 a, β^-) → ^{210}Bi (5 d, β^-) → ^{210}Po (138 d, α) → ^{206}Pb.

The radionuclides ^{238}U and ^{234}U have been used to evaluate the chemical weathering rate in hydrographic basins based on the fact that ^{234}U is preferentially mobilized to ^{238}U when rock weathers in the alteration front (Ivanovich and Harmon, 1992). Measurements of the $^{234}U/^{238}U$ activity ratio (AR) in rocks, soils and waters have allowed the calculation of the solution coefficient for the uranium characteristic of the region, that allows evaluate the time necessary to weather 1 m of rock under actual climatic conditions (Moreira-Nordemann, 1980, 1984; Conceição and Bonotto, 2003; Bonotto et al., 2007).

Table 13. Comparison of chemical analyses of the residual rock from Itararé Subgroup and its weathering products in Cr and BC horizons

Oxide	Residual Rock/Cr (w = 0.49)			Residual Rock/BC (w = 0.34)		
	Amount remaining (g)	Gain (+) or Loss (-) (g)	Gain (+) or Loss(-) (%)	Amount remaining (g)	Gain (+) or Loss(-) (g)	Gain (+) or Loss(-) (%)
SiO_2	41.600	-40.630	-49.4	26.790	-55.44	-67.4
Al_2O_3	2.820	0	0	2.820	0	0
CaO	0.025	-0.043	-63.2	0.030	-0.04	-54.4
MgO	0.060	+0.006	+11.1	0.060	+0.01	+19.1
Na_2O	0.033	-0.040	-54.8	0.026	-0.05	-65.0
K_2O	0.150	-0.292	-66.1	0.110	-0.33	-75.1
Fe_2O_3	0.600	+0.158	+35.7	0.860	+0.41	+93.5
MnO	0.004	0	0	0.004	0	0
TiO_2	0.170	+0.124	+270	0.190	+0.15	+322
P_2O_5	0.008	+0.001	+14.3	0.007	0	0
LOI	3.300	-10.510	-76.1	2.76	-11.10	-80.5
Sum	48.77	-51.22		33.66	-66.31	

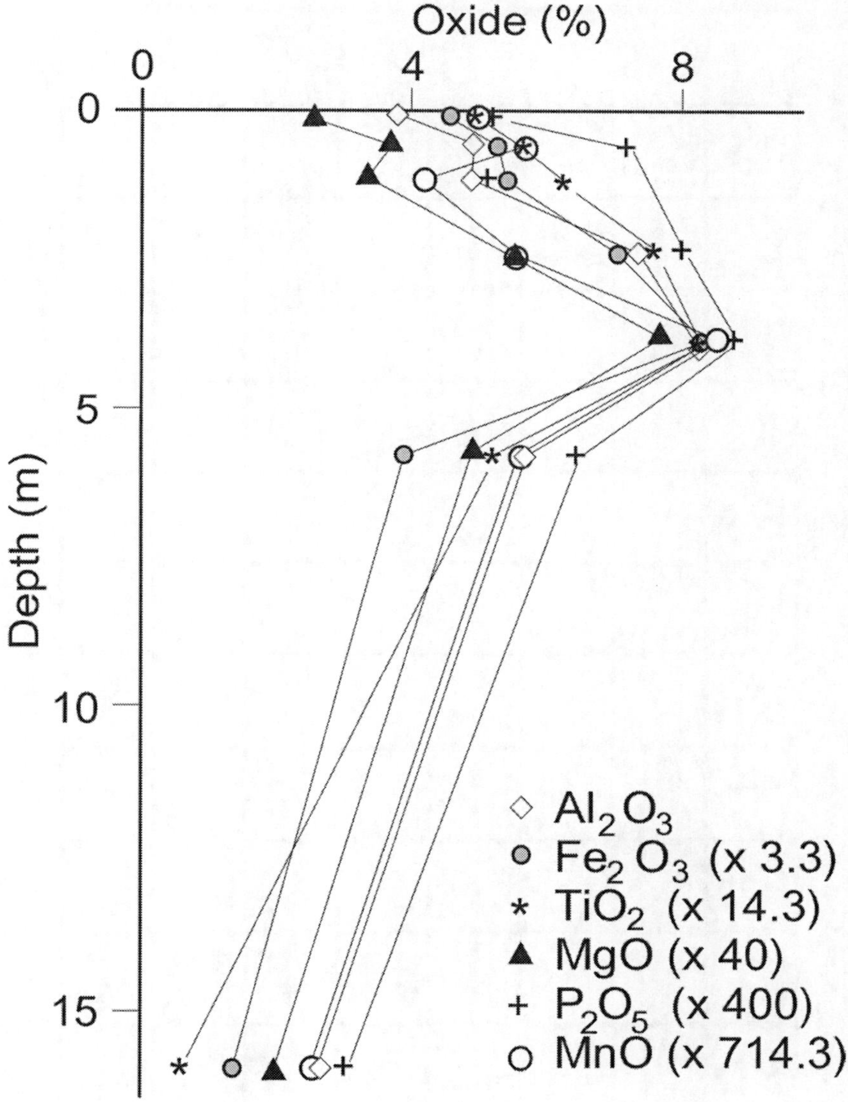

Figure 7. Chemical data of the soil profile developed over sandstone from Itararé Subgroup, São Paulo State, Brazil, plotted against depth.

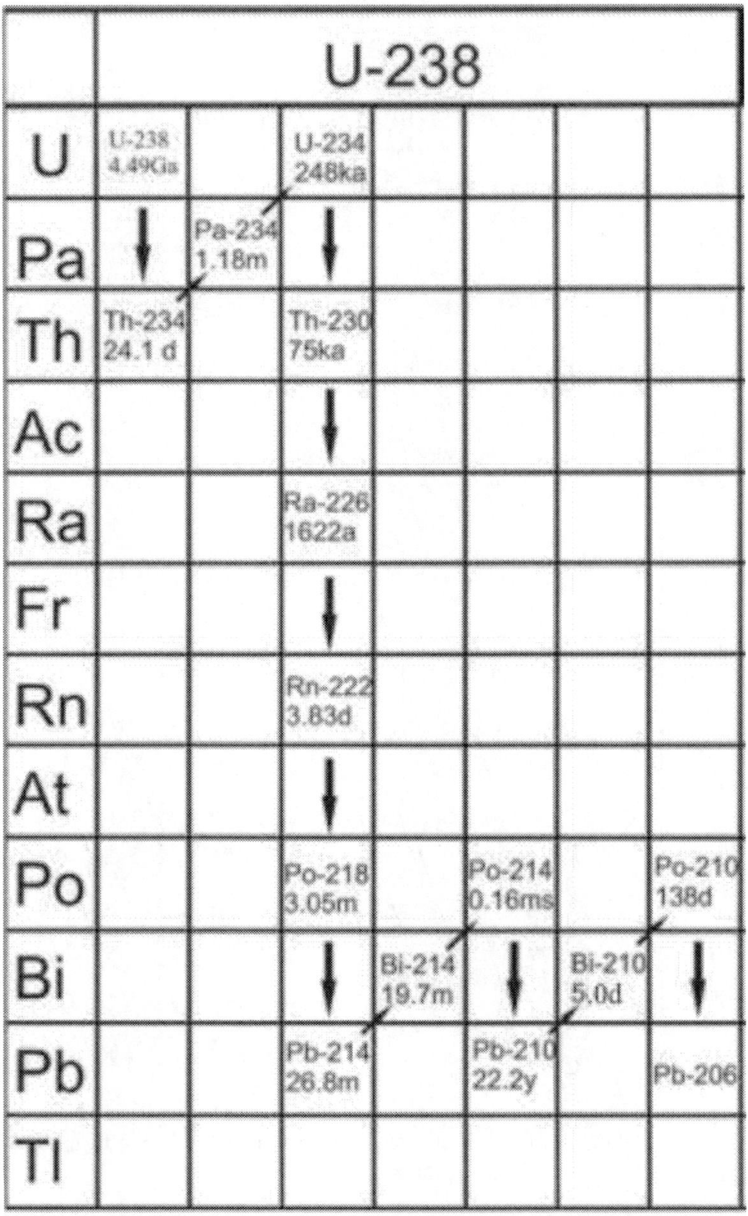

Modified from Ku (1976).

Figure 8. The natural uranium decay series.

Table 14 reports the results of the measurements of U content and AR realized by alpha spectrometry in the same samples given in Table 12. They had been completely dissolved in a mixture of HNO_3+HF+$HClO_4$ followed by a radiochemical procedure for extraction and electrodeposition of uranium isotopes on stainless-steel planchets (Veselsky, 1974; Osmond and Cowart, 1976; Ivanovich and Harmon, 1992). Uranium does not accumulate in the soil horizons associated with Al_2O_3, MgO, P_2O_5, Fe_2O_3, MnO, and TiO_2, since no significant correlation was found among their concentrations. Organic matter has often be considered an important complexing agent for uranium (Szalay, 1964), however, in the studied soil profile the correlation between these parameters also is not significant (r = 0.64). As expected, the data obtained for organic matter and loss on ignition correlate reasonably (r = 0.82), with the organic matter tending to decrease with increasing depth in the soil horizons (r = - 0.85) (Figure 9). The same trend was found for uranium (r = -0.86) (Figure 9), despite of the non-existence of correlation between organic matter and uranium.

Table 14. U content and $^{234}U/^{238}U$ activity ratio of the residual rock from Itararé Subgroup, São Paulo State, Brazil, and respective soil horizons

Horizon Code	U content[1] (μg/g)	$^{234}U/^{238}U$ A.R.
Ap	7.60	1.06±0.10
Bo1	6.71	0.97±0.09
Bo2	9.10	1.06±0.11
Bo3	7.88	1.17±0.18
BC	2.31	0.80±0.08
Cr	1.19	0.87±0.08
Residual Rock	1.40	0.90±0.07

[1] Uncertainty ±10% corresponding to 1σ standard deviation.

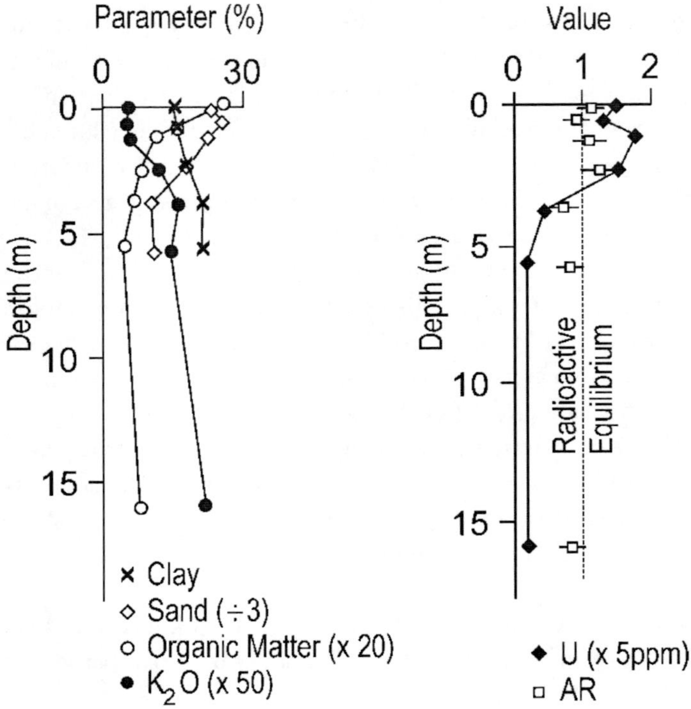

Figure 9. Chemical and U-isotopes data of the soil profile developed over sandstone from Itararé Subgroup, São Paulo State, Brazil, plotted against depth.

A possible explanation for such discrepant behavior is the different particle size distribution along the soil profile, also caused by influence of physical factors like humidity and temperature in addition to chemical processes acting there. Thus, there is a decrease of the sand fraction with increasing depth ($r = -0.92$), whereas an increase of the clay fraction with increasing depth ($r = 0.93$) (Figure 9). As expected, K_2O correlates very well with the amount of clay in the soil horizons ($r = 1$), and, consequently, K_2O content also increases with increasing depth ($r = 0.93$) (Figure 9). In contrast, U content decreases with increasing K_2O content ($r = -0.88$), as well with the increase of the clay fraction ($r = -0.84$). Uranium only correlated directly with the amount of sand fraction ($r = 0.82$) that increases towards the more superficial soil horizons. This suggests that it is contained in resistates/minor refractory minerals (such as zircon) which are highly resistant to weathering, so that they are incorporated intact in the coarser grained fraction of the soil horizons.

The four more superficial soil horizons exhibit ARs corresponding to unit, within experimental errors, indicating radioactive equilibrium between ^{234}U and ^{238}U in the detritic matrix at least over the last one million years (Latham and Schwarcz, 1987). Thus, there is no occurrence at these sites of extra ^{234}U-loss either due to α-recoil effects (Kigoshi, 1971) or by preferential leaching/etching of recoil-damaged sites (Fleischer, 1975). However, preferential removal of ^{234}U has occurred at the slightly altered parent rock, and Cr and BC soil horizons, generating ARs<1, within experimental errors, a typical result related to the effect of rock/soil-water interaction (Osmond and Cowart, 1976). These samples are located in the saturated zone of the soil profile, since the water table is between 3 and 4 m depth, situation that favors the production of ARs>1 for uranium dissolved in the liquid phase, and, consequently, ARs<1 in the solid phase.

Thus, non-significant U-dissolution from resistates by infiltrating more oxidizing rainwater is probably verified for samples taken from near the surface, yielding ARs=1, within experimental errors. The oxidation of U to the more soluble, hexavalent state (Langmuir, 1978) favors the preferential removal of ^{234}U by infiltrating water at present days for the deeper samples. The three deepest samples have ARs that are out of equilibrium, within experimental errors. They may supply timing informations concerning to the water-weathered profile interactions in the area. These disequilibria allow calculate an "apparent time" representing the state of disequilibrium acted against by the decay back to equilibrium of the shorter-lived daughter. This time (t) may be expressed for relative daughter deficiency by $t = 1/\lambda \ ln\ (1 - D/P)$ where λ is the decay constant of the daughter (^{234}U in the present case) and D/P is the activity of a daughter (D) to its parent (P) ($^{234}U/^{238}U$ in the present case) (Latham and Scharcz, 1986). The time needed to restore the system to equilibrium from deficiency may be calculated by the use of the quantity $5HF - t$ (HF is the half-life of ^{234}U) (Latham and Scharcz, 1986). Thus, taking into account the extreme disequilibria values of 0.8 and 0.9, it is possible to obtain the range of 410-660 ka as the time to decay back to equilibrium, which may represent the elapsed time for developing the more superficial (Bo1, Bo2, and Bo3) soil horizons during weathering of the sandstone parent rock.

CONCLUSION

Because the interaction of rainwaters, the atmosphere and the continental crust controls the chemical evolution of the Earth's surface, and, consequently, of the hydrographic basins, a lot of recent studies have been realized to determine the parameters controlling the denudation, chemical weathering and physical erosion under different climatic conditions. In order to estimate chemical weathering fluxes affecting the drainage basins, models utilizing the Na, Ca, K, Mg and total dissolved load concentrations have been applied to obtain fractions coming from dissolved rocks, after corrections for river waters of the atmospheric inputs, dissolved solids from ion exchange sites in clays minerals, dissolved solids due to changes in biomass and anthropogenic inputs, among other factors. Since uranium is among the main elements contributing to the natural terrestrial radioactivity, ^{238}U and its descendant ^{234}U of the mass number (4n+2) decay series have also been utilized in studies focusing hydrological resources in drainage basins. Significant relationship has been observed between the uranium content and the amount of sand fraction in soil horizons developed over sandstone from Itararé Subgroup, Paraná sedimentary basin, Brazil, that increases towards the more superficial soil horizons. $^{234}U/^{238}U$ disequilibria values of 0.8 and 0.9 found at the saturated zone have allowed suggest times between 410 and 660 ka for the development of the more superficial soil horizons during weathering of the sandstone parent rock. Therefore, the natural U-isotopes ^{234}U and ^{238}U are helpful tools for investigating weathering processes occurring in soil profiles.

ACKNOWLEDGMENTS

FAPESP (Foundation Supporting Research in São Paulo State) and CNPq (National Council for Scientific and Technologic Development) in Brazil are greatly thanked for financial support of related investigations described in this chapter.

REFERENCES

Allman, M.; Lawrence, D. F. Geological Laboratory Techniques; Arco Publishing Company, Inc.: New York, NY, 1972; 335 pp.

Almeida, F. F. M. de; Melo, M. S. In Mapa geológico do Estado de São Paulo; Pró-Minério; Ed.; PROMOCET: São Paulo, 1981; v. 1, pp. 46-81.

Amaral, G.; Cordani, U. G.; Kawashita, K.; Reynolds, J. H. *Geochim Cosmochim Acta*, 1966, 30, 159-189.

Araújo, L. M.; França, A. B.; Potter, P. E. *Hydrogeol J.* 1999, 7, 317-336.

Bagolini, B. *Mem. Mus. Tudentino di Sci. Nat. Ireuts*, 1971, 19, 71-122.

Balasubramaniam, K. S.; Moorthy, V. K.; Vyas, B. R. In *Proc. II Int. Seminar on Laterisation Processes*; 1983; pp. 577-590.

Baskaran, M. *J Geophys Res.* 1995, 100, 2833-2840.

Bertin, L. The new Larousse Encyclopedia of the Earth; Crown Publishers, Inc.: New York, 1974; 425 pp.

Bonotto, D. M.; Fujimori, K.; Moreira-Nordemann, L.M. *Appl Radiat Isotopes*, 2007, 65, 474-481.

Bowen, N. L. *J Geol.* 1922, 30, 177-198.

Colman, S. M. *US Geol Surv Prof Paper*, 1982, 1246, 1-51.

Conceição, F. T.; Bonotto, D. M. Environ Geol 2003, 44, 408-418.

DAEE (Departamento de Águas e Energia Elétrica) Estudo de águas subterrâneas - Região Administrativa 5 - Campinas; Secretaria de Obras e Meio Ambiente: São Paulo, 1981; 639 pp.

Engelhardt, W. von Chem Erde 1937, 11, 17-37.

Faure, G. Principles and applications of inorganic geochemistry; Mac Millan Publishing Co: New York, 1991; 626 pp.

Federici, P. R.; Pappalardo, M. *Rend Fis Acc Lincei.* 2001, 9, 69-82.

Fleischer, R. L. *Health Phys.* 1975, 29, 69-73.

Goldich, S. S. *J Geol.* 1938, 46, 17-58.

Humbert, R. P.; Marshall, C. E. *Missouri Agr Expt Sta Research Bull.* 1943, 359.

Ivanovich, M.; Harmon, R. S., Uranium Series Disequilibrium: Applications to Environmental Problems; Clarendon Press: Oxford, 1992; 2nd edn.

Jackson, M. L.; Sherman, G. D. Adv in Agron. 1953, 5, 219-318.

Kiehl, E. J. Interpretation of soil properties; Teaching text; ESALQ (University of Agriculture "Luiz de Queirós", USP-University of São Paulo: Piracicaba, SP, 1977; 237 pp.

Kigoshi, K. *Science*, 1971, 173, 47-48.

Ku, T. L. An Rev Earth Plan Sci 1976, 4, 347-379.

Langmuir, D. Geochim Cosmochim Acta 1978, 42, 547-569.

Laporte, L. F. Ambientes antigos de sedimentação; Ed. Edgard Blücher: São Paulo, 1969; 145 pp.

Latham, A. G.; Schwarcz, H. P. 1986. Models of uranium etching and leaching: their applicability to estimating rates of natural uranium removal from crystalline igneous rocks using U-series disequilibrium data from the Eye-Dashwa Lakes pluton; Canadian Atomic Energy, Rep. TR-353, 1986; 60 pp.

Latham, A. G.; Schwarcz, H. P. *Appl Geochem*. 1987, 2, 55-65.

Melfi, A. J. *Geochim Cosmochim Acta*, 1967, 31, 1079-1089.

Mitchell, R. H.; Bergman, S. C. Petrology of lamproites; Plenum Press: New York, 1991; 451 pp.

Moreira-Nordemann, L. M. *Geochim Cosmochim Acta*, 1980, 44, 103-108.

Moreira-Nordemann, L. M. *J Hydrol*. 1984, 71, 131-147.

Netto, A. S. T. Manual de sedimentologia; Petrobrás/SEPES/DIVEN: Salvador, 1980; 194 pp.

Szalay, A. *Geochim Cosmochim Acta*, 1964, 28, 1605-1614.

Osmond, J. K.; Cowart , J. B. *At Energy Rev*, 1976, 14, 621-679.

Pedro, G. *Rev Geogr Phys Géol Dyn*. 1968, X (5), 457-470.

Pedro, G.; Sieffermann, G. Review in modern problems of geochemistry; UNESCO: Paris, 1979; pp. 39-55.

Pettijohn, F. J. *J Geol*. 1941, 49, 610-625.

Poldervaart, A. *Geol Soc America Spec Paper*, 1955, 62, 119-144.

Renne, P. R.; Ernesto, M.; Pacca, I. G.; Coe, R.S.; Glen, J. M.; Prévot, M.; Perrin, M. *Science*, 1992, 258, 975-979.

Riebe, C. S.; Kirchner, J. W.; Granger, D. F.; Finkel, R. C. *Geology*, 2000, 28, 803-806.

Riebe, C. S.; Kirchner, J. W.; Finkel, R. C. *Geochim Cosmochim Acta*, 2003, 67, 4411-4427.

Riebe, C. S.; Kirchner, J. W.; Finkel, R. C. *Earth Planet Sci Lett*. 2004, 224, 547-562.

Schellmann, W. In *Proc. I Int. Seminar on Laterisation Processes*; 1979; pp. 1-10.

Small, E. F.; Anderson, R. S.; Hancock, G. S. *Geomorphology*, 1999, 27, 131-150.

Turner, S.; Regelous, M.; Kelley, S.; Hawkesworth, C. J.; Mantovani, M. S. M. *Earth Planet Sci Lett*. 1994, 121, 333-348.

UNESP (Universidade Estadual Paulista Júlio de Mesquita Filho) Petrologia Magmática: Caderno de Exercícios de Litoquímica. IGCE (Instituto de Geociências e Ciências Exatas): Rio Claro, 1980; 42 pp.

Veselsky, J. *Radiochim Acta*, 1974, 21, 151-154.

Waal, A.; Jiménez-Rueda, J. R.; Bonotto, D. M.; Bertelli, C.; Hoffmann, H. M.; Fosshag, E.; Santilli, M. In Environmental Health Risk V; Brebbia, C. A.; Ed.; WIT Press: Southampton, 2009; pp. 113-124.

Wentworth, C. K. *Jour Geol.* 1922, 30, 377-392.

Weyl, R. V. Z Pflanzenernähr Düng Bodenk 1952, 75, 135-141.

White, I. D. Relatório final da Comissão de Estudos das Minas de Carvão de Pedra do Brasil; Imprensa Nacional: Rio de Janeiro, 1908.

In: Weathering: Types, Processes and Effects ISBN 978-1-61324-280-3
Editor: Matthew J. Colonl pp. 167-188

Chapter 4

WEATHERING OF DIMENSIONAL GRANITIC STONES USED AS CLADDING

Patricia Vazquez and F. Javier Alonso
University of Oviedo, Department of Geology, Oviedo, Spain

1. INTRODUCTION

Granites have been used as building stones for centuries since they are very durable materials and allow different finishes such as polishing, flaming, hammering and so on. When they are placed in a building, their surface is exposed to a wide range of weathering agents, which can produce damage, and consequently changes, in their aspect. The resultant decay depends on their petrographic and physical characteristics, and also on the environment (Del Monte et al., 1996; Chabas y Jeanette, 2001; Esbert et al., 2004). Weathering forms in historical granite buildings are well known, such as detachment, discolouration, grain disaggregation, crusts formation and/or biological colonisation. However, weathering forms of new panels in existing buildings are less identified.

Rainwater or groundwater may enter the granite through fissures in the stone, usually by capillary rise. This may produce dissolution or chemical variations in the minerals (Suzuki et al., 1995). Subsequently, if the temperature decreases to below zero degrees, this water turns into ice, which produces tension on the walls of the fissures and causes the cracks to enlarge (Watts, 1985; Iñigo et al., 2000; del Río, 2005; Takarli et al., 2008). If this water contains salts, crystallisation may happen under common temperature and humidity conditions. Some salts, such as sodium sulphate, can increase their volume by more than 300% when they precipitate. This phenomenon leads to high pressures within the crack system and results in growth of the cracks (Schaffer, 1932; Winkler and Singer, 1972; Doehne, 2002; Alonso et al., 2008). Humidity in the stones, added to changes in temperature, may produce variations in the volume and shape of the stones. The bowing phenomenon in marbles, due to simultaneous temperature and humidity changes, is well known (Koch y Siegesmund, 2002; Scheffzük et al., 2004; Malaga et al., 2008). Moreover, this effect was also observed in granitic stones (Siegesmund et al., 2008; Vázquez et al., 2010). Pollution also plays a role in granite decay since "...soiling processes in urban environments are responsible for the build-up of unsightly dark deposits (patinas) with a complex inorganic/organic chemical composition on the surfaces of granitic buildings" (Schiavon et al., 1995).

Prior to undertaking a study of the durability of any material, it is necessary to know the properties that influence its decay. Hydric properties condition water uptake (Mosquera et al., 2000; Pender et al., 2004); dynamic properties provide insight into the crack system of the granites and whether changes are produced inside the stone (Wyllie et al., 1956; Sousa et al., 2005; Chaki et al., 2008). Thermal properties give an idea about the behaviour of the stones in response to temperature changes, which are also related to the mineralogy and texture of the stones (Castro de Lima y Paraguassú, 2004; Gómez-Heras et al., 2006). Surface properties such as colour, gloss and roughness are very important in an ornamental stone therefore it is essential to keep these aspects intact throughout the years. These surface properties are strongly related to the artificial finish applied to the natural stones (Huang et al., 2002; Benavente at al., 2003; Pinheiro Sousa, 2007).

In this study, four fresh granites commonly used as ornamental stones and with differences in their petrography are studied. The main objective is to determine the weathering forms of these ornamental stones when exposed to weathering agents such as salts, ice, pollution, and temperature and humidity variations. To do that, an initial stone characterisation is required, against which it is possible to assess the granites after weathering to determine variations in their bulk and surface properties. From the results, the petrography and the changes in the granite properties are related to their durability.

2. Materials

The granites were selected according to their different grain size and colour. They are GA (grey, medium grain size), G (grey, coarse grain size), RP (red, coarse grain size) and R (pink, porphiritic granite).

2.1. Methodology

2.1.1. Petrography

A petrographic description is carried out at both macro and micro scales. Mineral proportion and grain size were determined with Corel Draw tools in 10 x 10 x 2 cm slabs. GA and G were coloured with sodic cobaltonitrite to distinguish between feldspars. Streckeisen's (1976) classification was applied to the mineral proportion results. A microscopic description was carried out to complete the mineralogical characterisation with details such as anorthite content or accessory minerals. Chemical composition was obtained by X-ray fluorescence.

2.1.2. Bulk Properties

To assess the void system, density and porosity were determined. Apparent density (ρ) and open porosity (n) were calculated by water absorption techniques (UNE-EN 1936:1999). A further three elemental properties, related to water transport within the stone, were calculated according to the following standards: Free water absorption (UNE-EN 1936:1999), free water desorption (ICR-CNR NORMAL 29/88), and capillarity absorption (UNE – EN 1925: 1999). Four cubic specimens were tested with dimensions 5 x 5 x 5 cm. The velocity of p-wave propagation (Vp) was determined with a Pundit plus device with 1MHz transducers. The samples were measured in dry conditions. A quality index (IQ) was also determined following Tourenq et al., (1971) as follows: IQ= Vp/Vp* x 100 where Vp* is the theorical Vp calculated with mineral proportion and mineral Vp, and considering the stone without porosity. The linear thermal expansion coefficient was obtained by means of a Dil-6 device. The samples taken from each of the granites were 6 cylinders with 1.5 (diameter) x 5 (height) cms. The temperature interval was 20ºC to 90ºC. This cycle was repeated 5 times and the average of the resultant length increments was taken as the linear thermal expansion coefficient (α).

2.1.3. Surface Properties

To determine the surface properties of colour, gloss and roughness, four slabs of each of the granites, with dimensions 10 x 20 x 2 cm and polished surfaces, were measured. Colour was measured with a Minolta CR-200 colorimeter equipped with a pulse xenon lamp emitting an 8-mm diameter diffuse beam of illuminant C (average daylight) and 2° observer. Due to the heterogeneity of the materials, 100 shots were necessary in every slab. The standard CIE L*a*b* colour system was used. Three parameters were obtained, L* (lightness with values from 0 to 100), C* (chroma from 0 to 60) and h* (hue from 0 to 360). Gloss was measured with a Minolta glossmeter 60, which emits the light with an angle of 60º. 30 measurements were required in each slab, and the result was the average of four slabs, measured in gloss units (GU %). Surface roughness was measured using a Mitutoyo Surftest SV-2000N2 contact profilometer, fitted with a 2 μm diamond tip stylus, to which a load of 0.75 mN was applied to keep the speed constant and the tip in contact with the surface at all times. A micrometric table attached to the roughness tester ensured that the same profiles were taken before and after the tests. Preliminary studies were carried out in order to determine the best measurement and evaluation characteristics (Table 1) (Alonso et al, 2007; Alonso et al., 2008; Vázquez Menéndez et al., 2008). The parameters selected to evaluate the roughness were: Ra: Arithmetical mean deviation of the profile; Rp: Maximum profile peak height; Rv: Maximum profile valley depth, and Ry: Maximum height of the profile.

2.2. Characterisation

The results of the characterisation of the four granites are summarised in Tables 1 to 4.

In Table 1 the main petrographic parameters are summarised. GA is a two mica, grey, medium grain sized and equigranular monzogranite with oligoclase (An_{12}). G is a coarse, heterogeneous monzogranite with grey alkali feldspar and oligoclase (An_{15}). RP is a coarse, heterogeneous sienogranite with red alkali feldspar and oligoclase (An_{11}). R is a porphyritic quarztsienite, with pink alkali feldspar megacrystals up to 60 mm and oligoclase (An_{24}). Chemical composition is similar in all of the granites, although it is pertinent to note that G owns the highest proportion of Ca and Fe, the chemical compounds more susceptible to weathering. Porosity is similar among the granites, at slightly higher than 1%.

Table 1. Main characteristics of the studied granites

Name	Grain size (mm)			Composition (%)					Apparent density Kg/m³		Apparent Porosity (%)	
	x	σ		F	P	Q	M		x	σ	x	σ
GA	4	1	Medium	37	23	25	15	Monzogranite	2630	1	1.2	0.04
G	11	4	Coarse	35	32	25	8	Monzogranite	2635	3	0.7	0.03
RP	15	6	Coarse	50	14	30	6	Syenogranite	2605	3	1.1	0.05
R	25	7	Very Coarse	60	20	12	8	Quarztsyenite	2605	3	0.9	0.02

F: alkali feldspar; P: plagioclase; Q: quartz; M: mica; x: average; **σ** standard deviation. The grain size was divided as follows:

<1mm	*Fine*
1mm to 5mm	*Medium*
5mm to 30 mm	*Coarse*
> 30mm	*Very coarse*

In Table 2 the main coefficients resulting from the granites′ hydric properties are summarised. All the results indicate that GA has the highest porosity and fastest water movement kinetics, while G the lowest. The other two granites have intermediate values. Table 3 shows velocity of p-waves propagation with similar relations between granites than in hydric properties, and thermal linear expansion coefficient, which shows the lowest values in RP and the highest in G, both with similar grain size.

Table 2. Hydric, dynamic and thermal parameters. Basic and hydric properties

	%Ws	Absorption %S			Desorption %S			C (g/m²sg)	
		1h	1d	3w	1h	1d	3w	x	σ
GA	0.45	83	96	96	58	32	1	1.9	0.2
G	0.26	67	85	84	50	33	9	0.4	0.1
RP	0.41	67	89	92	61	46	20	0.7	0.1
R	0.33	76	89	100	63	45	17	1.2	0.1

% Ws: water content in saturation conditions. %S: Saturation degree in percentage.1h:1 hour; 1d: 1 day; 3w: 3 weeks; C: capillarity coefficient;. x: average; σ standard deviation

Table 3. Dynamic and thermal parameters. Basic and hydric properties

	Vp (m/s)			α (10^{-6}/ºC)	
	x	σ	IQ	x	σ
GA	4420	5	77	7.8	0.3
G	5420	50	91	9.8	0.8
RP	5325	65	91	7.6	0.7
R	5125	200	88	8.6	1.4

Vp: velocity of p-waves propagation; α: thermal lineal expansion coefficient;. x: average; σ standard deviation

Table 4 shows the measured parameters of the surface properties i.e. colour, gloss and roughness. In spite of the evident colour differences among the granites at visual scale, the measurements registered very similar chromatic parameters (greyish). Gloss may vary in relation to the grain size and the hardness of each mineral. In polished granites values are between 60 and 80%, and are usually related to roughness. Roughness parameters are quite similar in all four granites because of their polish finish. The valleys (pores and cracks) tend to determine the maximum roughness value because they are about 4 to 5 times deeper than the peaks are high. The process of polishing the granite erodes the surface but does leave some biotite lamina as peaks above the flat surface; meanwhile the cracks below the surface are not affected by the polishing process.

Table 4. Surface properties: colour, gloss and roughness

Name	Colour				Gloss	Roughness (μm)			
	L*	C*	h*	Visual Colour	(% GU)	Ra	Rp	Rv	Ry
GA	65	5	110	Pale Grey	70	1	6	22	28
G	60	4	140	Grey	65	1	5.5	25	30
RP	60	11	80	reddish	75	1.2	4	30	35
R	70	6	110	Pale pink	60	1.2	5.5	30	35

L*: Lightness; C*: chroma; h*: hue; Ra: average roughness; Rp: peak roughness; Rv: valleys roughness; Ry: maximum roughness.

3. Durability

The durability of the stones was assessed by means of artificial ageing, such as salt crystallisation, freeze-thaw cycles and SO_2 exposure. The effect of temperature and humidity changes was assessed by means of a bowing test. The salt crystallisation and freeze-thaw tests were modified in order to force the exposure of the stone surface to the salt and ice.

3.1. Methodology

3.1.1. Salt Crystallisation Test

The Salt Crystallisation test was carried out in two different ways. Both of them follow the standard UNE-EN 12370:1999 with immersion of the stones in a 14% sodium sulphate solution followed by drying. The first test followed the standard stated above, where the dry stage was carried out at 105ºC in cubic specimens to allow the salt to crystallise within the stone. Due to the low porosity of the granites the test was extended for 60 cycles. The damage was evaluated by means of variations in Vp and capillary uptake. The second test was developed with modifications in the dimensions (10 x 10 x 2 cm with polished surface) and with the dry stage conducted at laboratory conditions (20ºC and 50% relative humidity), to force the migration of the salt and its crystallisation as efflorescence. In this test, the damage was focused on the surface and the variations were assessed by surface properties.

3.1.2. Freeze-Thaw Test

The Freeze Thaw Test was carried out based on the standard UNE-EN 12371:2002 in which the stones are frozen immersed in water and thawed out of it under laboratory conditions. Del Río et al. (1995), Wessman (1997), Iñigo et al. (2000) and Takarli et al. (2008) all studied the evolution of the crack network and related properties of granitic stones when exposed to freeze-thaw cycles and their results showed that for low porosity granites hardly any variations exist. Due to that, in this study the specimens were immersed during the whole test, i.e. for both the freeze and unfreeze stages, to force the damage onto the surface as a result of ice formation and melting. The test was performed for 200 cycles, where each cycle consisted of 6 hours freezing at -20ºC and 6 hours thawing at +20ºC. At the end of the test, another cycle of continuous freezing over 25 days was performed. Bulk properties (Vp and capillary uptake) were measured at every 50 cycles on 5 x 5 x 5 cm cubes with one polished surface; gloss and roughness were measured at the end of all the tests.

3.1.3. SO_2 Exposure Test

The SO_2 exposure test was carried out following the standard UNE-EN 13919:2003 but also with some modifications. The test consisted of placing the specimens in a closed recipient with sulphurous acidic solution for a period of three weeks to create an acidic atmosphere. The samples were of two types: 12 x 6 x 1 cm slabs with polished surface to measure surface properties, and smaller specimens for microscopic evaluation and chemical analysis. Two tests were carried out with different temperatures:

- 20ºC and 350ml H_2SO_3 + 150ml H_20 for 50 l recipe
- 60ºC and 350ml H_2SO_3 + 150ml H_20 for 50 l recipe

3.1.4. Bowing Test

The Bowing Test was conducted according to the Nordtest Method NT BUILD 499 (2002). Three slabs of each granite were placed on a bed of grit with a distance of 50 mm to the heating system. The slab dimensions were 40 x 10 x 1.5 cm because the current trends in building stone manufacturing are to produce thinner slabs. In addition, Siegesmund et al. (2008) observed a higher degree of bowing in thinner slabs.

The test performed included 36 heating-cooling cycles divided into four stages of nine repetitive cycles with varying temperatures and humidity. The heating took 6 hours followed by 18 hours of cooling at room temperature:

- Stage 1: heating up to 90ºC in dry conditions
- Stage 2: heating up to 90ºC in wet conditions
- Stage 3: heating up to 120ºC in dry conditions
- Stage 4: heating up to 120ºC in wet conditions

The wet conditions were obtained by adding water to the grit bed to get a wet bottom surface and a dry upper surface (Siegesmund et al., 2008).

The bowing was measured after each cycle with a measurement bridge (accuracy ± 1 µm / 35 cm). The zero position was calibrated using a core of silica glass with coplanar surfaces. The deformation was detected in mm/m, calculated according to the circle equation of Koch and Siegesmund (2004), where positive values indicate convex bowing and negative values indicate concave bowing.

3.2. Results

3.2.1. Weathering within the Stone

The tests showed that both salt crystallisation and freeze-thaw cycles may act within the stone to modify the crack network and consequently may affect the bulk properties of the granite. Table 5 shows the Vp and C variations expressed as a percentage at the end of the test in relation to sound stone. Salt crystallisation can be seen to result in much more damage than ice formation; the damage is even evident with the naked eye.

Table 5. Variations in Velocity of p-wave propagation (Vp) and Capillarity coefficient (C) before and after salt crystallisation test and freeze-thaw test expressed in percentage

	Salt crystallisation test				Freeze-thaw test			
	ΔVp (%)		ΔC (%)		ΔVp (%)		ΔC (%)	
	x	σ	x	σ	x	σ	x	σ
GA	-9	2	1	17	-3	2	9	14
G	-33	1	910	360	-2	3	66	44
RP	-29	2	560	61	-3	6	67	82
R	-20	4	124	14	-3	3	9	24

x: average; σ standard deviation

Salt crystallisation test: visually, only slight damage was observed after 45 cycles. The damage was similar in all the granites (mica detachment, crack growth and micro-cavities formation in plagioclase), however the intensity of the damage varied. During 45 cycles, the Vp decreased lineally in the three less porous granites, from values around 5000 m/s down to values around 4000 m/s, which was similar to the initial values of GA. In the last 15 cycles, the Vp did not change, and the Vp values of all the granites were similar, despite the initial values. Changes in capillary uptake were more evident, with initial values around 1 $g/m^2s^{1/2}$ and final values up to 5 $g/m^2s^{1/2}$ in RP and about 4 $g/m^2s^{1/2}$ in G. The decrease in Vp and the huge increase in capillary uptake were evidence of the crack opening and growth with salt crystallisation. Only GA, the most cracked stone, exhibited a very slight change in Vp and no variation in capillarity uptake.

Freeze-thaw test: the bulk properties showed the same trend as the salt crystallisation test, although much weaker. Vp tended to decrease but the final variation was almost imperceptible. C increased only in the two, more cracked granites with an increase slightly higher than 50 %.

3.2.2. Weathering on the Stone Surface

Surface properties experienced more evident changes if related to bulk properties. Colour, gloss and roughness varied in all the tests carried out although with different intensity in every test.

Salt crystallisation: colour hardly showed any variation, with only a slight decrease in lightness at the end of the test in RP (Table 6). The decreases in gloss at the end of the test (indicated in Table 6) may have been due to an increase in roughness focused on peaks (Vazquez, 2010). Only in two granites (RP and R) was the loss of gloss progressive throughout the cycles. Roughness variations are also shown in Table 6. After salt crystallisation, the peak roughness increased because the mica exhibited greater reliefs than the rest of the minerals, thus the mica were responsible for the increase in peak roughness. Peak values after the test reached 35-40 µm and GA (the two mica granite) had the highest peaks due to its higher mica content. Valley roughness also increased after the test because of the cracks and grain boundaries opening, from 20-30 µm in the sound stones to 55-70 µm after the test. Feldspars remained almost flat (Figure 1). SEM images affirmed the damages observed with the naked eye. Figure 2 shows a plagioclase with Ca-rich core after alteration, mica lamina opened and detached and loss of material in alkali feldspar.

Table 6. Variations in surface properties after salt crystallisation test. Colour and gloss are expressed in absolute values and roughness in percentage

Name	Colour			Gloss (%GU)	Roughness			
	ΔL*	ΔC*	Δh*		ΔRa	ΔRp	ΔRv	ΔRy
GA	-1.2	0.2	-4.2	-12	100	300	450	250
G	-1.7	0.1	0.7	-32	60	130	490	70
RP	-3.0*	-0.1	2.2	-25	90	250	870	150
R	-1.3*	1.3*	-3.3	-11	150	420	640	340

L*:Lightness; C*: chroma; h*: hue; Ra: average roughness; Rp: peak roughness; Rv: valleys roughness; Ry: maximum roughness.

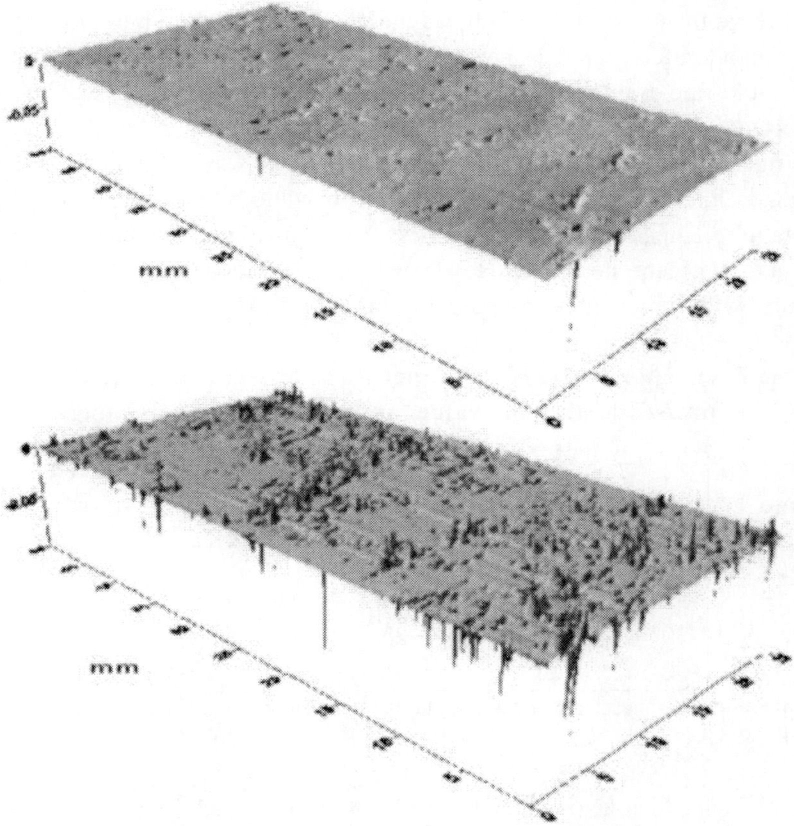

Figure 1. 3D representations of G before the salt crystallisation test (top) and after 30 cycles (bottom). Mica peaks are clearly observed while alkali feldspars remain flat.

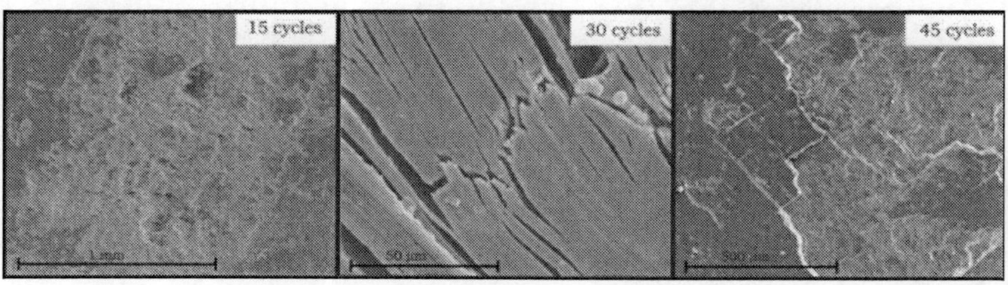

Figure 2. SEM images of the granites in the salt crystallisation test. Left: Weathered plagioclase with the Ca-rich core showing more alteration than the extremities. Centre: Biotite with a crack perpendicular to the exfoliation planes. Right: Loss of material in alkali feldspar.

Freeze thaw test: Table 7 shows the variations of gloss measured in GU and roughness measured in percentages. Gloss suffered a decrease from 17 GU in the fine grain size granite to almost 60 GU in G. However, the most important damage was not the loss of glossiness in itself but the heterogeneity of the loss. Aesthetically, the surface was deteriorated because it looked like it had small gloss spots on it, while others appeared opaque. Average roughness and peak roughness hardly exhibited any change, only R had a significant decrease. However, valleys (cracks) closed in all the granites with final values around 5-20 μm.

Table 7. Variations in surface properties after freeze-thaw test. Gloss is expressed in absolute values and roughness in percentage

Name	ΔGloss (%GU)	Roughness			
		ΔRa	ΔRp	ΔRv	ΔRy
GA	-17	-15	-15	-75	-70
G	-55	2	41	-70	-60
RP	-30	-27	-1	-75	-70
R	-50	-45	-50	-70	-70

Ra: average roughness; Rp: peak roughness; Rv: valleys roughness; Ry: maximum roughness.

SO_2 exposure test: when the slabs were removed from the recipient, crystals were observed on the surface. The slabs were cleaned and their colour, gloss and roughness were measured. Visually a change in colour to yellow was easy to observe and some minerals had turned to shades of orange. Table 8 shows the variations in surface properties in these tests. Colour changes were very remarkable in grey granites, showing a decrease in lightness, increase in chroma and an evident and homogeneous yellowing measured as a decrease in hue. In some cases, the yellowing was not so homogeneous since orange spots were observed on some slabs. Pink granites showed an increase in the chroma measurement, although with the naked eye the variations were not very evident (Figure 3). Gloss exhibited a decrease in all the granites of about 30% in relation to its original values. Roughness showed the same trend as demonstrated in the salt crystallisation test, but with less intensity, with an increase in peaks due to the mica detachment. This effect was more evident in grey granites at low temperature and in pink granites at high temperature. Under SEM, the specimens were observed without washing in order to observe the different crystals which would crystallise on the surface and to analyse their composition (Figure 4). The main characteristics found by microscopic evaluation were that there were a lot of crystal habits in all the tested specimens (Figure 4 left and right), but they were more developed in low temperature tests (at 20ºC gypsum crystals were rather prismatic and at 60ºC rather acicular). Crystals did not appear in all minerals equally, they tended to crystallise in Ca rich minerals or in those with higher relief (Figure 4 centre). There were crystals with a lot of impurities which did not show the typical gypsum habit.

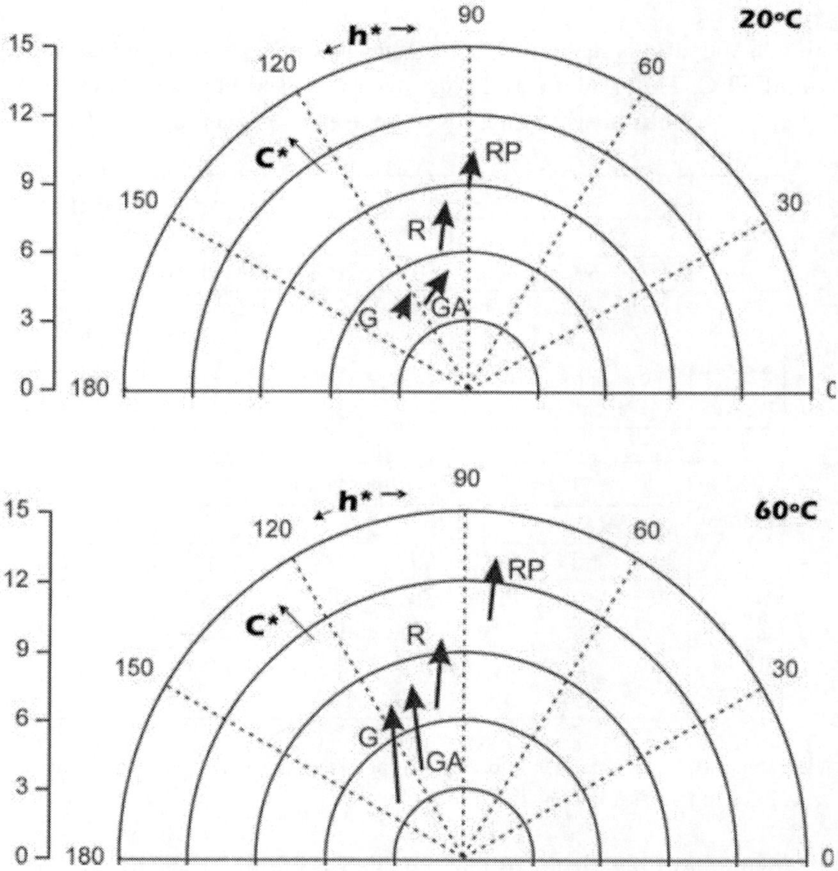

Figure 3. Colour variations in SO_2 test. Top: At 20ºC there is a variation in C* and h*. Bottom: At 60ºC h* variation is stronger in the grey granites while C* increases in all the granites with similar magnitude.

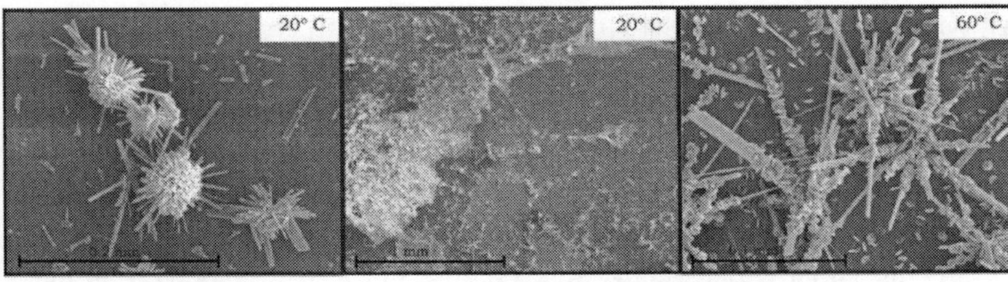

Figure 4. SEM images of the granites after SO_2 exposition. Left: Example of gypsum crystals observed at 20°C. Centre: Crystallisation occurs in higher relief areas such as mica or grain boundaries. Right: Different crystal habits with acicular morphology at 60°C.

Table 8. Variations in surface properties after SO_2 exposure test at 20°C and 60°C. Top: Colour and gloss are expressed in absolute values. Bottom: roughness is expressed in percentage.

Name	Colour						Gloss (%GU)	
	ΔL*		ΔC*		Δh*			
	20°C	60°C	20°C	60°C	20°C	60°C	20°C	60°C
GA	-1.7	-3.1	0.9	3.5	-12	-6.2	-25	-20
G	-1.0	-1.6	0.4	3.4	-6.8	-20	-10	-20
RP	-0.6	-0.4	1.3	2.5	-1.9	0.4	-30	-25
R	0.1	0.8	2.6	-3.3	-3.8	2.8	-3	-15

Name	Roughness (μm)							
	ΔRa		ΔRp		ΔRv		ΔRy	
	20°C	60°C	20°C	60°C	20°C	60°C	20°C	60°C
GA	-45	-10	75	35	-80	-3	-70	0
G	1	-5	110	70	-70	-5	-40	0
RP	-10	10	0	175	6	20	6	30
R	7	-20	35	100	25	-4	30	4

L*:Lightness; C*: chroma; h*: hue; Ra: average roughness; Rp: peak roughness; Rv: valleys roughness; Ry: maximum roughness.

Bowing test: the results of the bowing test in three slabs of each granite are given in Figure 5, with the outcome of each third cycle shown. During the *first stage*, with heating up to 90°C and dry conditions, the granites hardly exhibited any deformation. G showed the greatest concave bowing, between

0.3 and 0.35 mm/m. In the *second stage*, with the same temperature but adding water to the bottom surface, the trend in deformation varied. GA became concave in the first cycle of the stage, but thereafter, the general deformation of all the stones was to convex shapes. Only R displayed remarkable behaviour, by reaching values up to 0.7 mm/m, which was significantly higher than the values measured in the rest of the stones. In the *third stage*, with the increase in temperature and dry conditions, almost all the samples showed concave or no further deformation. In the *fourth stage*, all the slabs bent in a convex direction and the deformation values for most of them were comparable to those shown throughout the earlier stages. However, three of the granites exhibited considerable bowing. G and RP showed similar bowing in all the slabs, up to 1 and 0.7 mm/m respectively. R was the granite that experienced the highest bowing of up to 3 mm/m at the end of the test. At the end of the test clay minerals were observed filling cracks in the upper face.

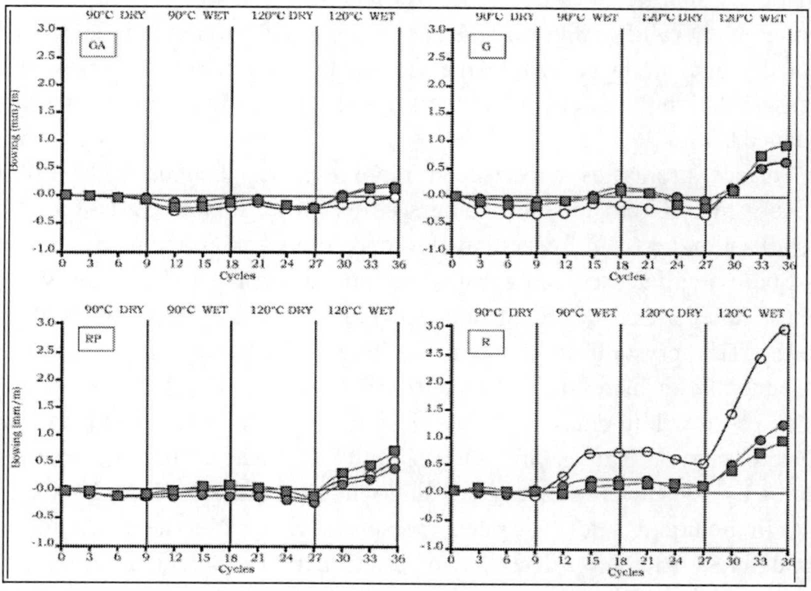

Figure 5. Bowing test performed on each slab from the four granites. The highest bowing belongs to the R' slab with oriented alkali feldspars.

4. DISCUSSION

All the studied granites suffered slight weathering after the artificial ageing tests and the decay patterns were quite similar among the different weathering agents.

In relation to the bulk granite, velocity of p-wave propagation and capillarity coefficient were the parameters used to assess the variations in the stone crack system due to their close relationship to one another (Vázquez et al. 2010a). In both tests, the damage measured was very slight. A decrease in velocity of p-waves, and an increase in capillarity coefficient, means crack opening and the appearance of new-formation cracks. An increase in cracks also means that in the next salt or ice event, the damage will occur in more locations throughout the stone. The damage produced was generally slight in both tests, but was greater in salt crystallisation than in the freeze-thaw test. The decrease in velocity of p-waves can even be classified as ′not statistically significant′ after the freeze-thaw test. The low porosity of all the granites with values around 1% leads to the conclusion that in spite of the great damage produced within the cracks, the volume of the cracks is so small in relation to the volume of the stone that the suffered decay was hardly evidenced.

Surface properties experienced more evident changes with different intensity in different tests; roughness suffered the highest variation in salt crystallisation, gloss in freeze-thaw test and colour in SO_2 exposure.

Sodium sulfate, ice and gypsum crystallise on the surface, mainly in the highest relief areas, which in polished granites belong to mica exfoliation planes. The crystallisation produces mica planes on the surface and consequently an increase in the peaks of roughness. Crystal formation also leads to a growth in cracks and the formation of micro-cavities, which can be translated into an increase in valleys (Figure 1). Sodium sulfate increases its volume by 300 times when crystallising while ice only increases by 4% over water in the liquid state. This phenomenon correlates with the higher increase in roughness after the salt effect, medium increase in roughness after SO_2 exposure and gypsum crystal formation, and almost no change after the ice damage. Increase in roughness in itself is not a great damage, visually the change can be considered as non-aesthetically changed. However it must be taken into account since an increase in roughness facilitates the adhesion of airborne particulates with can lead to soiling and even crust formation. Rough surfaces are also optimal for moss, lichen and other bio-organisms to create colonies, which may result in further damage and/or non-aesthetic changes.

Gloss is a very important surface property when a granite with a polished finish is studied. A decrease in gloss was observed in all of the three tests, however the intensity of the decrease differed. As mentioned above, sodium sulfate damage was focused mainly on higher relief areas so the loss of gloss was also produced around those areas. Mica planes and cracks are areas with low gloss so the general loss of gloss seen in these areas was the lowest of all the three tests. The SO_2 exposure test created an acidic atmosphere which acted on the whole surface, but also with higher intensity in high relief areas, similar to the salt crystallisation test. SO_2 also produced a chemical attack on the surface. Both factors acting together resulted in a higher loss of gloss than demonstrated by salt crystallisation. The highest loss of gloss was observed in the freeze-thaw test due mainly to the effect of the ice on the whole surface. The main characteristic of this damage was that the loss of gloss exhibited an irregular pattern. These heterogeneities were more evident in very coarse minerals, such as alkali feldspars in R. On the contrary, GA exhibited an evident loss of gloss measured with the glossimeter, although visually this granite hardly showed the loss of gloss due to its smaller grain size and segmented grain boundaries.

Colour changes are related mainly to chemical damage. Working with sound granites, in the freeze-thaw test no colour changes were expected and none were observed. In the salt crystallisation test colour variations were only significant in pink granites at the end of the test with a decrease in lightness and an increase in chroma. The decrease in lightness may have resulted from some white plagioclase in the pink granites which darkened to brownish colours. An increase in chroma means more vivid colours, which can be explained by the mica planes on the surface. After saw cutting, mica planes exhibit a grey colour, however after salt crystallisation their relief changes and the mica planes exhibit an intense brown colour. After SO_2 exposure, colour changes were remarkable in grey granites (GA and G), showing a yellowing, which became more intense with higher temperature. Even orange spots could be observed on some slabs. The pink granites (RP and R) showed an increase in red and yellow colours in the measurements, although with the naked eye the variations were not very evident (Figure 3). The chemical attack produced a yellowing of the surface, which in grey granites meant a general yellowing of the stone and in red-pink granites simply a change in hue to yellow, which is not evident to the human eye. The weathering produced by SO_2 in granites must be taken into account in grey-white granites since the colour change can be observed with the naked eye and the final results may not be the expected ones. Pink-red granites do not exhibit any variation that is perceptible to humans, therefore they are considered optimal for use in polluted areas.

Bowing of granitic stones was observed under high temperatures and extreme humidity conditions. Although these conditions cannot be found in real environments, after a long time of exposure stones may keep residual strains and bowing may happen. In this case it was essential to know which characteristics of the stones had special influence on deformation. In the case of the bowing test in the four selected granites, grain size, orientation and mineral composition together seemed to be the most influencing parameters. The smaller and more homogeneous granite (GA) did not show any deformation, while the porphyritic granite (R) showed bowing that could be observed with the naked eye. The two intermediate grain size granites showed intermediate deformation (Figure 5). The porphyritic granite had about 70% of feldspar, the thermal expansion coefficient of which is the lowest among minerals. However, 60% of them were very coarse grained and had the greatest bowing when they were oriented with the long axis parallel to the long side of the slabs (*note:* slab dimensions were 40 x 10 x 1.5 cm). Clays may also have played a role in the deformation since they appeared concentrated in the cracks on the surface of R, the porphyritic granite (Vazquez et al., 2010b).

5. CONCLUSION

All the granites studied are very durable materials in relation to the weathering agents tested. The bulk effects were minimal, thus their strength remained almost constant before and after weathering.

Since these granites have similar porosity and initial alteration, their grain size seems to be the most important characteristic in influencing the deterioration of the stone.

All the studied granites have a low porosity, close to 1%. Due to that, the damage which can be produced within the stone, by salt or ice formation, is hardly evident. Salt produces higher damage than ice, but the damage is very low in both cases.

Surface properties measured after salt crystallisation, freeze-thaw and SO_2 exposure tests showed evident changes. Roughness changes were most evident after salt crystallisation, gloss changes after the freeze-thaw test and colour changes after SO_2 exposure.

Colour variations after SO_2 exposure were evident with the naked eye only in grey granites. In both the pink granites after SO_2 exposure, and in all the granites after salt crystallisation, some colour changes were produced but they could only be confirmed after measurement with the colorimeter.

Gloss decreased after all three tests (salt crystallisation, freeze-thaw and SO_2 exposure). The freeze-thaw test produced a non-homogeneous loss of gloss resulting in a non-aesthetic aspect. This kind of weathering form was more evident in polished surfaces and in coarse grain granites.

Roughness varied in all the tests in which it was measured. It did not result in aesthetic changes; nevertheless it may have contributed to the deposition of pollutants, airborne salts and even the growth of bio-organisms.

Bowing of granitic stones was only observed with extreme temperatures and humidity, and was related to mineralogy, grain size and grain orientation. In this research, the granite with the highest grain size, and those slabs with orientated megacrystals, showed the highest bowing.

REFERENCES

Alonso, F. J., Vázquez, P., Esbert R. M., Ordaz, J. (2007). "Influence of measuring conditions on roughness parameters of ornamental rocks." Workshop: Preservation of Natural Stone and Rock Weathering, Taylor & Francis . 13-16.

Alonso, F. J., Vázquez, P., Esbert, R. M., Ordaz, J., (2008). "Ornamental granite durability: evaluation of damage caused by salt crystallization test." *Materiales de Construcción*, 58(289-290): 191-201.

Benavente, D., Martïnez-Verdú, F., Bernabeu, A., Viqueira, V., Fort, R., García del Cura, M.A., Illueca, C., Ordöñez, S., (2003). "Influence of surface roughness on color changes in building stones." *COLOR research and application*, 28(5): 343-351.

Castro de Lima, J. J., Paraguassú, A.B., (2004). "Linear thermal expansion of granitic rocks: influence of apparent porosity, grain size and quartz content." *Bulletin of Engineering and Geological Environment*, 63: 215-220.

Chabas, A., Jeannette, D., (2001). "Weathering of marbles and granites in marine enviroenment: petrophysical properties and special role of atmospheric salts." *Environmental Geology*, 40(3): 359-368.

Chaki, S., Takarli, M., Agbodjan, W.P. (2008). "Influence of thermal damage on physical properties of a granite rock: Porosity, permeability and ultrasonic wave evolutions." *Construction and Building Materials*, 22(7): 1456-1461.

Del Monte, M., Ratazzi, A., Romao, P., Rossi, P., (1996). "The role of lichens in the weathering of granitic Buildings." Degradation and Conservation of granitic rocks in monuments. *Environmental protection and conservtaion of the European cultural Heritage.*: 301-306.

del Río, L. M., López, F., Esteban, F.J., Tejado, J.J., Mota, M., González, I., Ramos, A., San Emeterio, J.L., (2005). "Ultrasonic Study of alteration processes in Granites caused by freezing and thawing." *IEEE Ultrasonics Symposium*: 415-418.

Doehne, E. (2002). "Salt weathering: a selective review." Natural Stone, Weathering Phenomena, Conservation Strategies and Case Studies. Ed. Siegesmund S., Weiss T., Vollbrecht A., Geological Society, London, Special Publications. 205: 51-64.

Esbert, R. M. (2004). "Deterioro del granito en los monumentos de Santiago." 2º Congreso internacional sobre la conservación del granito, Santiago de Compostela.

Gómez-Heras, M., Smith, B.J., Fort, R., (2006). "Surface temperature differences between minerals in crystalline rocks: Implications for granular disaggregation of granites through thermal fatigue." *Geomorphology*, 78: 236-249.

Huang, H., Li, Y., Shen, J.Y., Zhu, H.M., Xu, X.P., (2002). "Micro-structure detection of a glossy granite surface machined by the grinding process." Journal of Materials Processing Technology 129: 403-407.

ICR CNR NORMAL 29/88 (1988). "Misura dell Indice de Asciugamento, Roma."

Iñigo, A., Vicente MA, Rives V (2000). "Wheathering and decay of granitic rocks: its relation with their pore network." *Mechanical of materials*, 32: 555-560.

Koch, A., Siegesmund, S., (2002). "Bowing of marble panels: on-site damage analysis from the Oeconomicum Building at Göttingen (Germany)." Natural Stone, Weathering Phenomena, Conservation Strategies and Case Studies. Ed. Siegesmund S., Weiss T., Vollbrecht A., Geological Society, London, Special Publications. 205(299-314).

Koch, A., Siegesmund, S (2004). "The combined effect of moisture and temperatura on the anomalous expansion behaviour of marble."

Environmental Geology, 46: 350-363.

Malaga, K., Schouenborg, B., Grelk, B., (2008). "Bowing and expansion of natural stone pannels." *Materiales de Construccion*, 58(289-290): 97-112.

Mosquera, M. J., Rivas, T., Prieto, B., Silva, B., (2000). "Capillary Rise in Granitic Rocks: Interpretation of Kinetics on the Basis of Pore Structure." *Journal of Colloid and Interface Science*, 222: 41-45.

NT BUILD 499:2002 Cladding panels: Test for bowing.

Pender, R. J. (2004). "The behaviour of water in porous building materials and structures." *Reviews in Conservation*, 5: 49-62.

Pinheiro Sousa, F. J., Júnior, N.V., Weingaertner, W.L., Alarcón, E., (2007). "Glossiness distribution over the surface of stoneware floor tiles due to the polishing process." *Journal of Material Science*, 42: 10124-10132.

Schaffer, R. J. (1932). "The weathering of natural building stones." Department of scientific and industrial research, london. Special Report, 18.

Scheffzük, C., Siegesmund, S., Nikolayev, D.I., Hoffmann, A., (2002). "Texture, spatial and orientation dependence of internal strains in marble: a key to understanding the bowing of marble panels?" Natural Stone, Weathering Phenomena, Conservation Strategies and Case Studies. Ed. Siegesmund S., Weiss T., Vollbrecht A., Geological Society, London, Special Publications. 205: 237-249.

Schiavon N, C. G., Schiavon G, Fabbri D (1995). "Nature and decay effects of urban soiling on granitic building stones." *The Science of the Total Environment*, 167 87-101.

Siegesmund, S., Rüdrich, J., Koch, A., (2008). "Marble bowing: comparative studies of three different public building facades." *Environmental Geology*, 56: 473–494.

Sousa, L. M. O., Suárez del Río, L.M., Calleja, L., Ruiz de Argandoña, V.G., Rodríguez Rey, A., (2005). "Influence of microfractures and porosity on the physico-mechanical properties and weathering of ornamental granites." *Engineering Geology*, 77: 153–168

Streckeisen (1976). "To each plutonic rocks its proper name." *Herat Science*, 12: 73-85.

Suzuki, K. O., Kuwahara, M.T., Hirama, K., (1995). "Material porperty changes in granitic rock during long term immersion in hot water." *Engineering Geology*, 40: 29-39.

Takarli, M., Prince, W., Siddique, R., (2008). "Damage in granite ynder heating-cooling cycles and water freeze-thaw condition." *International*

Journal of Rock Mechanics and Mining Sciences, 45: 1164-1175.

Tourenq, C., Fourmaintraux, D., Denis, A., (1971). "Propagation des ondes et discontinuités des roches." Symposium de le Societé Internationelle des Mechániques des Roches, Nancy: 15 pp.

UNE-EN 1925:1999 Determinación del coeficiente de absorción de agua por capilaridad.

UNE-EN 1936 (1999). Métodos de ensayo para la piedra natural. Determinación de la densidad real y aparente y de la porosidad abierta y total.

UNE-EN 12370:1999 Métodos de ensayo para la piedra natural. Determinación de la resistencia a la cristalización de sales .

UNE-EN 12371:2002 Determinación de la resistencia a la heladicidad.

UNE-EN 13919-2003 Determinación de la resistencia al envejecimiento por la acción del SO2 en presencia de humedad.

Vázquez, P., Siegesmund, S., Alonso, F.J. (2010). "Bowing of dimensional granitic stones." Environmental earth sciences (In Press).

Vázquez, P. (2010). " Ornamental Granites: Characterisation, Durability and Suggestions of use." Doctoral Thesis. University of Oviedo.

Vázquez, P., Alonso, F.J., Esbert, R.M., Ordaz, J. (2010). "Ornamental granites: Relationships between P-Wave velocity, water capillary absorption and crack network." *Construction and Building Materials*, 24(12): 2536-2541.

Vázquez-Menéndez, P., Esbert, R.M., Alonso, F.J., Ordaz, J., (2008). "Evaluation of damage induced by salt crystallization in granitic building stones." 11th International congress on Deterioration and Conservation of Stone, Torún. Ed Jadwiga W. Lukaszewicz y Piotr Niemcewicz. I: 325-332.

Watts, S. H. (1985). "A scanning electron microscope study of bedrock microfracturas in granites under high artic conditions." *Earth Surface Processes and Landforms*, 10: 161-172.

Wessman, L. (1997). "Studies on the frost resistance of natural stone." Report TVBM-3077; Licentiate Thesis, Lund University: 212.

Winkler, E. M., Singer, P.C., (1972). "Crystallization pressure of salts in stone and concrete." *Geological Society of America Bulletin*, 83: 3509-3514.

Wyllie, M. R. J., Gregory, A.R., Gardner, L.W., (1956). "Elastic wave velocities in heterogeneus and porous media." *Geophysics*, XXI(1): 41-70.

In: Weathering: Types, Processes and Effects ISBN 978-1-61324-280-3
Editor: Matthew J. Colonl pp. 189-206 © 2011 Nova Science Publishers, Inc.

Chapter 5

Desert Pavement Types and Development and Implications for Rates and Processes of Formation: Case Study of the Northern United Arab Emirates

Asma Al Farraj *

Geography Department, United Arab Emirates University, P.O. Box 17771, Al Ain, United Arab Emirates

Abstract

There are arguments in the literature about the desert pavement mechanism and processes. One argument suggests that desert pavements result from wind deflation, the other argues that they result from mechanical weathering and concurrent soil formation. In the northern United Arab Emirates (UAE) the two types of desert pavement are found. The desert pavements which develop by wind deflation are found in depressions between sand dunes, and desert pavements which develop by mechanical weathering and soil formation are found on fluvial surfaces (terraces and alluvial fans). The first type occurs as a result of the winnowing removal of fine material from the surface, exposing underlying gravels. In the UAE these are characterised by a

* E-mail: asma@uaeu.ac.ae

fine texture of small stones (cm size) protecting soil beneath. The second type involves mechanical weathering of clasts in situ at the same time as soil formation takes place. The stones are modified by weathering processes, which results in the fragmentation of the clasts. Windblown fines are trapped within the fabric and washed through the pavement surface. Wetting and drying and perhaps bioturbation reorganise the clasts to produce smooth interlocked pavement surfaces.

The rate of desert pavement development is different between the two types and may differ within the same type according to the weathering behaviour. For instance, for the deflation type, the deflation rate may be similar on surfaces of different age, producing pavement properties that may not be age-related. They can develop within hours (24-48), depending on the wind speed. For the mechanical weathering type, different lithologies or a different climatic regime may influence the rate of development, but generally the pavement properties (type of rock and weathering mechanism) will show clear age-related properties. On deflation pavements fragment angularity may be less well defined than on mechanical weathering pavements, and may shows a decrease in angularity on older pavements. Clasts on mechanical weathering pavements become progressively more angular with age and smaller in size in relation to the subsurface material. However, salt weathering and rock varnish may smooth the edges on older surfaces. Furthermore, microtopography on the pavement surface, related to original deposition becomes progressively obliterated. Age-related differentiation on both types of pavement may be undertaken through a consideration of soil maturity (colour and texture and profile development).

Key words: desert pavement, weathering, deflation.

1. Introduction

Desert pavement has been identified as armoured surfaces of closely packed stones (cm size) composed usually of angular to sub-rounded fragments, one or two stones thick, set on or in matrices of finer material usually soil (Springer, 1958; Cooke, 1970; Yaalon, 1970; Cooke and Warn, 1973; Al-Farraj, 1996; 2002; Al-Farraj and Harvey, 2000). Desert pavements are widespread in arid regions, where they have different ethnic names (e.g.; Saih in United Arab Emirates (UAE), Reg in Israel, Hamada in Morocco, Gibber plain in Australia and stone mantle/pavement in US). The degree of desert pavement development has been used as a tool for the correlation and relative dating of desert surfaces (McFadden, et al, 1989; Al-Farraj, 1996;

2008; Al-Farraj and Harvey, 2000, Guralnik, et. al., 2010). Cooke et. al. (1993) suggested that stone pavement controls surface stability and hydrology(Wainwright, et al, 1995; McDonald, et al, 1996), and records chemical and physical processes on desert surfaces, and serves as a source and storage for archaeological material (Cooke, et. al,1993).

There has been discussion in the literature about desert pavement initiation. Some studies in the Sahara and Arabia have suggested that they result from wind deflation. In these areas the desert pavement is remarkably level and smooth, together with the absence of fine material (Symons and Hemmings, 1968; McCauley, et al, 1979, Misak, 2000). However, studies in the US and UAE suggest that they "born on the surface" where they occur as a result of mechanical weathering in conjunction with dust accumulation and soil formation (McFadden, et al, 1987, 1998; Walles, 1995; Al-Farraj, 1996, Al-Farraj and Harvey, 2000, Al-Farraj, 2008). Similarly, in Israel desert pavement formation is interpreted as part of soil development processes (Amit and Gerson, 1986, Amit et al, 1993, Guralnik, et, al, 2010). Cooke et al, (1993) put forward three reasons to desert pavement formation; 1) wind deflation of fine materials, creating a coarse-grained lag. 2) Surface runoff and lateral transport by creep, resulting in the winnowing of fine sediment. 3) Shrink-swell processes of soil causing upward migration of clasts to the land surface (Cooke, 1970, Bread et al, 1997). However, He did not differentiate between the environment and other factors for each type.

This study suggests that there are actually two types of desert pavement; one resulting from winnowing of fine sediment from fluvial flash flood deposits terminated in depressions between sand dunes. The other type results from long-term physical

weathering on fluvial deposits (alluvial fans and terraces) within wadis and on alluvial fans at mountain fronts. For this case study of desert surfaces in northern UAE two different locations have been chosen; One in Wadi Lamha for the study of deflation pavement and the other at Saih Dibba for the study of physical weathering pavement (Figure 1). The studies of deflation involve the analysis of their characteristics and experimental monitoring of the formation of the deflation desert pavement. The study of weathering desert pavements was designed to identify the effect of different clast lithology on the clasts of the pavement, but not to identify age sequences. In other words, this study is based on desert pavement processes only; it does not consider age relations. For this reason the samples of the mechanical weathering pavements has been chosen from the same age surfaces according to Al-Farraj (2002).

2. PHYSICAL SETTING

Northern UAE has a wide range of geomorphology; including fluvial, aeolian and marine features. The fluvial environment includes wadis and gravel plains. On the terrace, fan and gravel plain surfaces there are desert pavements formed by weathering. Within the aoelian environments there is the other type of desert pavement formed by winnowing.

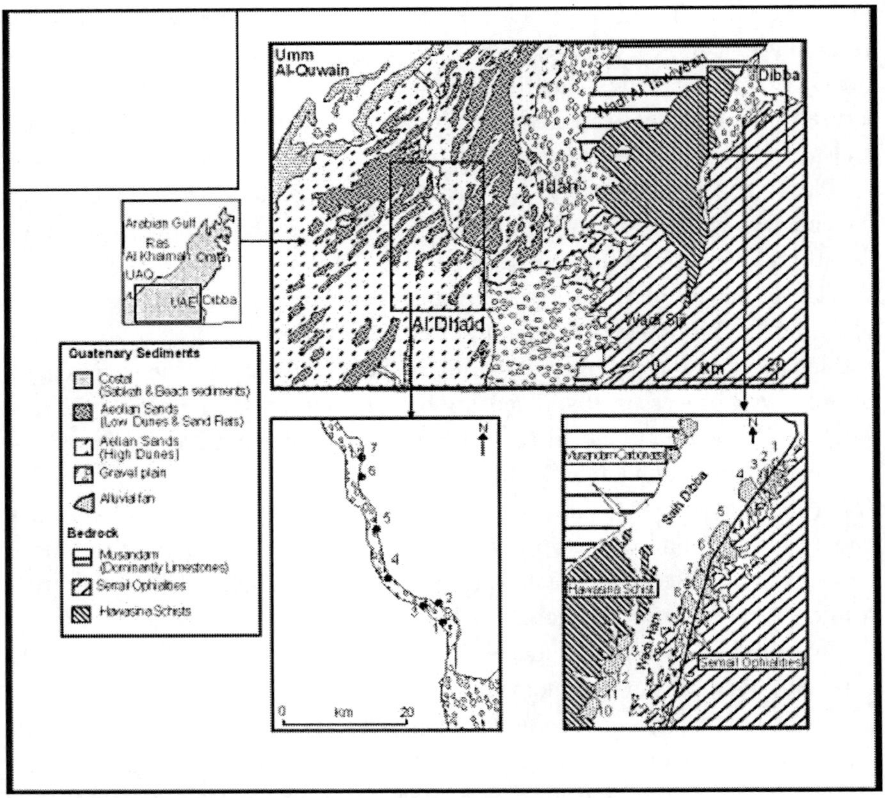

Source: The National Atlas of the United Arab Emirates, 1993.

Figure 1. Location of the study area and Quaternary geology. A) site locations of deflation desert pavements (in Wadi Lamhah). B) site locations of physical weathering pavements on Saih Dibba alluvial fans.

Wadi Lamha is on the western part of the Oman Mountains, fringing the mountains is the Al Dhaid gravel plain, on which physical weathering desert pavements are developed. Wadi Lamha crosses the gravel plain into the dune field (Figure 1). Wadi Lamha is the lower wadi of a group of wadis that drain west (Ashwani, Siji, Showka) and combine together at Falaj Al-Mualla from where Wadi Lamha extends for 20 km to the Arabian Gulf at Al Midfaq (Figure1). In the first 10 km only the wadi may flood between adjacent dunes. In this area winnowing desert pavement is common. Beyond this area the lower part of the wadi is floored by inland and coastal sabkha (Figure 1).

Saih Dibba is a gravel plain in the lower part of Wadi Ham (Figure 1). The Saih formed on a major fault zone in the northern UAE. These faults result in a cluster of three different types of rocks; igneous (Semail Ophiolites), sedimentary (Musandam carbonates) and Metamorphic (Hawasina schists). Several fans are fed from these different bedrocks; 9 fans from igneous rocks, 4 from metamorphic and 3 from sedimentary rocks (Figure 1). The desert pavement characteristics have been investigated on these fan surfaces, to demonstrate this desert pavement type development on all three rock types.

The modern climate in the study area is arid. Rain is irregular and infrequent; occurring mainly in winter. The average rainfall is 129 mm in Dibba and 120 mm in Falaj Al-Mualla. Mean winter (January) temperatures range from 17.5 0C to 20 0C. Mean summer (July) temperatures range from 34.8 0C to 36 0C.

3. METHODOLOGY

For the winnowing desert pavement, a survey has been carried out in Wadi Lamha at 7 locations (Figure 1). At all sites on the exposed surface 0.5 m2 quadrat samples of the pavement layer (maximum depth sampled 1-2 cm) were removed and taken for sieving in the laboratory, using the same methods as Symons and Hemming (1968). Soil texture and colour were recorded in the field to differentiate between exposed desert pavement surfaces. The sampled surfaces were at the side of the valley among the sand dunes at locations 1 and 2, and within the wadi course at the rest of locations. To measure the rate of deflation pavement development a square from 1 m2 quadrat was removed to 10 cm depth then mixed up and returned to the surface. The site was then re visited every day over period of 4 weeks (Fig 5 D). The rate of desert pavement formation was then related to wind speed records of the Ministry of Communications.

For the physical weathering desert pavement a survey had been carried out in Saih Dibba at 16 locations (Figure 1). In order to focus on the weathering processes differences and not age all surfaces have been chosen from moderately developed “C-age” fan according to Al-Farraj (2002), for this reason study sites were restricted to the same age the criteria used by Al-Farraj (2002). The exposed pavement surfaces reflect the rock types. At all sites scaled photographs of 1 m2 quadrats of the exposed surface were taken in the field. Soil texture and colour were recorded in the field. Measurement of clast size, angularity, and sorting on the pavement surfaces was carried out from the photographs. On each surface photograph clasts length and width of the largest 25 stones were measured, using the same method as Al-Farraj (1996) and Al-Farraj and Harvey (2000). Percentage cover was calculated. The percentage

was calculated by systematically sampling 100 points from a grid drawn onto the photograph and recording clast size at each sample point (Al-Farraj, 1996; Al-Farraj and Harvey, 2000). The angularity of the 25 stones was recorded using the Powers angularity chart (Tucker, 1996).

Table 1. Soil characteristics of location in Wadi Dhaid

Location	Soil colour	Soil description
1	10 YR 6\4 (light brown)	Silty\sand
2	10 YR 6\6 (reddish brown)	sandy
3a	10 YR 5\4 (brown)	Silty\sand
3b	10 YR 5\4 and 10 YR 4\4 (brown)	Sandy + black material
4a	10 YR 5\4 and 10 YR 4\4 (brown)	Loose black sandy material (no soil)
4b	10 YR 6\4 (light brown) and 10 YR 4\4 brown	Loose black sandy material (no soil)
4c	10 YR 5\3 brown)	Loose black sandy material (no soil)
5	10 YR 6\4 (light brown)	Loose black sandy material (no soil)
6	10 YR 6\4 (light brown)	Loose black sandy material (no soil)
7	10 YR 6\4 (light brown)	sandy material with mud at the surface

4. RESULTS

The results can be divided into 2 sets according to the desert pavement type; results for deflation desert pavements and results for physical weathering desert pavements. The results for deflation desert pavement surfaces resemble each other, and could not be differentiated in the basis of clast size or angularity. Therefore, the underlying soil has been used to differentiate between the surfaces (Table 1, Figure 2), to assess whether there appear to be age related difference between the sites. The soil is well developed at only 2 locations (1 and 3a Figure 2). Both are on stable and older surfaces, and both are located between sand dunes at the side of the Wadi. Most of other surfaces which are located within the wadi course have a less mature soil, which may indicate their modern occurrence.

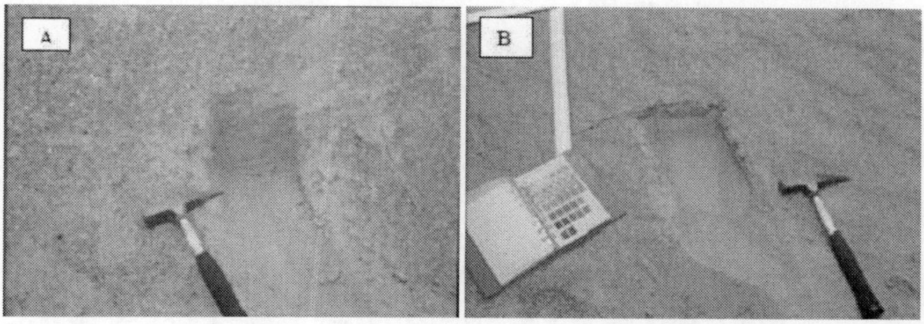

Figure 2. A) Soil development beneath deflation desert pavement. B) sediments beneath a lag layer of larger sand size (mm).

There are only small differences in sediment size, which is may be related to depositional processes, rather than to pavement weathering processes. Clast size tends to decrease down the wadi (Figures 3, 4).

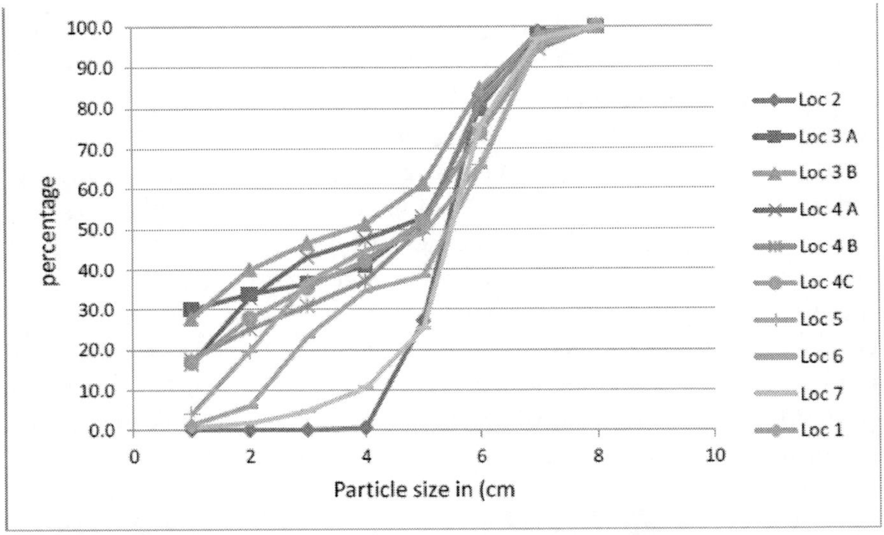

Figure 3. Clast size distribution along Wadi Lamhah.

Figure 4. Defilation pavement of different clast size on different sites locations at Wadi Lamhah, for clast size distribution see Figure 3.

Soil development beneath the surface may indicate time of deposition, and period of stability. It is might indicate that the sediments gets finer toward the distal wadi. The surface gets younger in age in locations distal from the mountain area. In the study area it may related to the tectonics of the area, as the wadi course has moved westward with time (Al-Farraj and Harvey, 2004).

Figure 5. A) Occurrence of deflation pavement on man-made mixed materials. B) Deflation pavement on lee sides of dunes where is the slope is not very steep. C) Lee sides of dunes with only sand size sediments. D) Deflation pavement in the experimental site, after 1 week. E&F) Development of deflation pavement on tyre marks and footprints.

On some low sand dune surfaces a (lag) layer of larger sized sand (mm-size) forms a surface similar to that on the surfaces in the depressions between the dunes. Location 4 shows that deflation surfaces may occur on any man-made surface where different size material has been mixed together (Figure 5D.). At this site material from the side of the wadi was used as a dam in the middle of the wadi to direct the wadi course away from the road. The wind has had the same effect on this material and developed a deflation desert pavement on the top surface of the dam and on the low angle sideslope of the dam. Furthermore, this indicates that deflation occurs not only on flat surfaces, but also on low angle surfaces. This is what has occurred at location 5, where deflation desert pavement has developed on the gentle slope of the lee side of a dune. The slip surface it has no desert pavement because it is too steep (Figure 5). Location 5 has mm-sized surface stones. The processes on the surface trace the patterns left on it (e.g.; tyre marks, footprints). This is may indicate that deflation occurs on only the top few centimetres of the surface not more. This go along with the result of the experiment which showed that the deflation pavement develops within 24 to 48 hours in wind speeds ranging between 20 and 10 knots.

In the physical weathering surfaces there are difference in pavement surfaces proprieties on each fan groups according to the rock type. However, all pavements are well sorted and interlocked (Figure 6 Table 2).

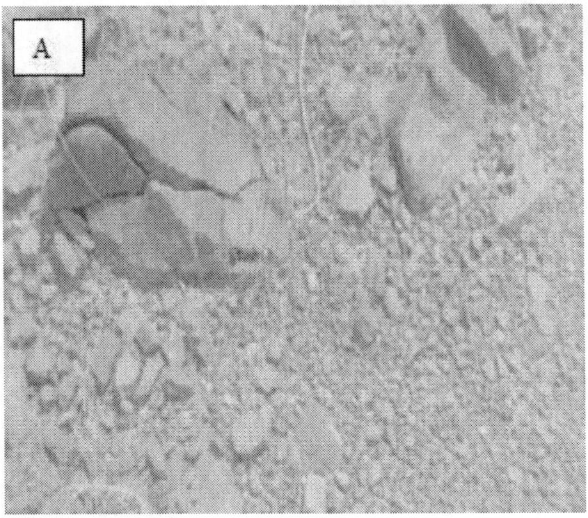

Figure 6. (Continued on next page).

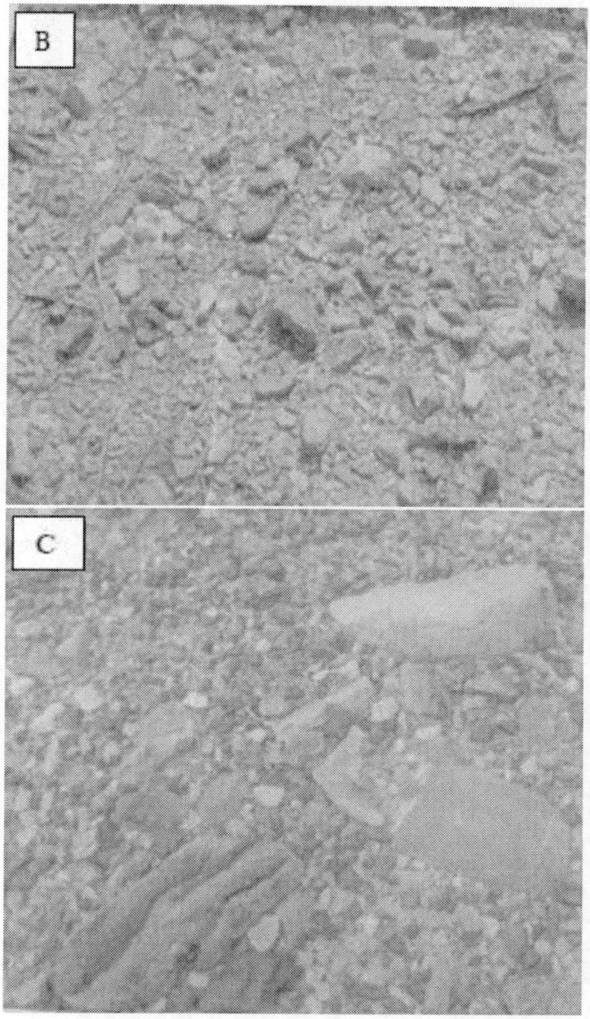

Figure 6. Physical weathering surface on different rock types. A) On sediments derived from igneous bedrock. B) On sediments derived from metamorphic bedrock materials. C) On sediments derived from limestone sedimentary bed rock.

Table 2. Desert pavement particle size characteristics, clast angularity, and pavement classifications derived from measurements made from photographed of 1-m2 samples at 12 selected C-age fan surfaces in Dibba

Bedrock type	Site	Largest 1-25 clasts a-axes (cm) a					Percentage cover by clasts with b-axis (cm) b				Clasts angularity: largest 25 clasts c							Pavement classification score d					PDIe
		1	2	5	10	25	>10	>7	>5	>4	VA	A	SA	SR	R	WR	Mnf	Clast size	Sort-ing	Angul-arity	Fract-ures	Surface texture	
Igneous rock	1	4.9	5.4	6.4	8.3	24.5	-	1	1	2	-	5	10	10	-	-	1.2	3	3	2	3	3	
	2	3.2	4.4	6.4	7.8	25.4	-	2	2	3	-	4	10	11	-	-	1.2	3	2.5	2.5	3	2.5	
	3	4.9	5.9	6.3	7.3	14	-	-	1	1	-	7	10	8			1.2	3	2.5	2	3	2.5	
	4	6.2	6.9	9.9	12.7	24	-	2	3	4	-	3	10	12	-	-	1.2	3	2	2	3	2.5	
	5	6.1	6.6	7.2	7.2	18	-	1	2	2	-	4	8	13	-	-	1.2	3	2	2	2.5	3	
	6	5.4	5.4	6.4	7.1	23	-	-	2	3	-	3	8	14	-	-	1.2	3	2	2	2.5	2	
	7	5.7	5.7	6.7	7.8	19	-	-	1	1	-	3	7	15	-	-	1.2	3	2	2	3	2.5	
	8	7	7.3	8.1	9.3	16	-	-	1	1	-	4	8	13	-	-	1.2	3	2	2	3	2.5	
	9	5	5.5	6.5	8.3	18	-	-	1	1	-	4	7	14	-	-	1.2	3	2	2	3	2.5	
Schist rock	10	2.8	3	3.4	4.3	55	-	1	1	2	2	11	10	1	-	-	1.2	3	3.5	3	2	3	
	11	3.1	1.9	4	4.6	9.5	-	-	-	2	2	10	10	3	-	-	1.2	3	3.5	3	2	3	
	12	1.9	2.1	2.5	2.7	9.2	-	-	-	1	3	11	9	2	-	-	1.2	3	3.5	3	2	3	
	13	1.8	3.7	2.3	2.8	18.7	-	-	-	1	3	12	8	2	-	-	1.2	3	3.5	3	2	3	

Bedrock type	Site	Largest 1-25 clasts a-axes (cm) a					Percentage cover by clasts with b-axis (cm) b				Clasts angularity: largest 25 clasts c							Pavement classification score d					PDIe
		1	2	5	10	25	>10	>7	>5	>4	VA	A	SA	SR	R	WR	Mnf	Clast size	Sort-ing	Angul-arity	Fract-ures	Surface texture	
Carbon-ate rock	14	3.1	3	4	4.6	19.5	-	-	1	1	2	12	10	1	-	-	1.2	3	3.5	3	2	2.5	
	15	3	3	4	4.7	18.9	-	-	1	2	3	12	10	-	-	-	0.6	3	3.5	3	2	3	
	16	4	4.3	5	5.9	244	-	1	2	3	-	6	12	7	-	-	1.2	3	3.5	3	2	2.5	

a a-axis (cm) for the 1st 2nd, 5th, 10th and 25th largest stone within the 1-m^2 sample quadrats.

b Percentage cover of the quadrates occupied by clasts > 4,5,7, and 10 cm b-axis.

c Clast angularity of the 25 largest clast, classified on powers roundness scale; VA= very angular, A=angular, SA= sub-angular, SR=sub-rounded, R= rounded, WR=well rounded.

d as suggested by Al-Farraj and Harvey, 2000, see the text for details.

e Pavement development index derived from the mean of the five categories (clast-size, sorting, angularity, fracturing, surface texture) above no pavement or immature pavement (0-1), moderately developed pavement (2-3), mature or well-developed pavement (3-4) for more details see Al-Farraj and Harvey 2000.

f average values derived from allocating a scale of VA=0 to WR=5

The results for physical weathering desert pavements are divided into 3 groups according to the rock type of the fan drainage basin. The results of the igneous rock fan group shows that they have relatively larger sediment size (between 3.2 and 25 cm a-axis), mainly sub-rounded to sub-angular stone fragments. Most of large stones have parallel cracks. The pavement is well sorted and well interlocked. Rock varnish makes the edges smother. The Schist rock fan group has smaller sediment size (between 3 and 15 cm a axis). The smaller size of sediments may be related to weak resistance of the schist rocks. It has more angular stone fragments; it is well sorted but not well interlocked. The carbonate rock fan group has larger sediments size (3.1 to 25 cm a-axis), roughly the same sediment size as on the igneous rocks and on pavements of the same age on Wadi Al-Bih fans (Al-Farraj, 1996). The sediment angularity ranges between angular and sub angular. The surface texture is well sorted and well interlocked.

5. Discussion

Putting together the literature on desert pavement formation mechanisms with the result of this study may suggest that the research on desert pavements was in different depositional environments. Studies in the Sahara and Arabia stress deflation pavements are in the same environment as Wadi Lamhah. Looking at the details of the study areas shows that in these areas dune sands and small stone fragments are common. Deflation pavements in Wadi Lamha occur only where aeolian dunes interact with fine fluvial gravels. Weathering studies in the US, Australia and Israel and previous studies of the UAE are mainly in fluvial environments, with limited or no dune development and gravel surfaces are common. This is similar to the cases described here, on gravel pavements in Saih Dibba that occur on coarse gravels on fluvial terraces and alluvial fans. This may suggest that both arguments may be right. In this study the two sites are less than 70 km apart. Elsewhere in the UAE the two pavement types often occur within 10 km of each other.

Deflation pavements take hours (24-28) to occur depending on the wind speed (20 to 10 knot). For this reason this type of desert pavement cannot be used to determine age relationships. Because all surfaces on this type of pavement look the same within hours. The sweeping of fine materials is only in first few centimetres, so it may obscure any pattern beneath it. It is may occur on flat and gentle slope surfaces. Thus, the levelling that had been argued in the literature does not occur from deflation, but may reflect the original depositional style.

The weathering desert pavement develops from physical weathering and for the same reasons suggested in the literature for mechanical weathering; salt, insulation and/or temperature change, and moisture weathering (see e.g. Goudie, 1997; Ollier, 1963; 1965). Salt weathering is common on salt lakes such as on Lake Eyre shoreline in South Australia (See Al-Farraj, 2008). Insolation may also apply (Al-Farraj, 1996; Al-Farraj and Harvey, 2000). Moisture and temperature change including freezing and thawing may apply in the Nevada desert (Cooke, 1993). In addition to mechanical weathering on fluvial surfaces is the accumulation of windblown dust and the development of the soil profile. It is well known that soil formation also involves break down of parent materials by mechanical and chemical weathering, and the vertical movement of the weathering products by water (Paton, et al, 1995, Amit and Gerson, 1986). Soil profiles on desert surfaces develop the mechanical weathering of clasts concurrent with by the accumulation of fines and the vertical movement of weathering products (McFadden, et al, 1987).

6. Conclusion

The landscape in northern UAE suggests that there are two types of desert pavement; resulting from two different processes; deflation desert pavement and physical weathering desert pavement. Deflation desert pavement results from winnowing of finer sediments on palaeo wadi sediments between sand dunes. It may occur on any mixed materials even man-made materials. It may occur on flat surfaces and on gently sloping surfaces as on dune lee faces. This type cannot be used as surface age indicators.

The Physical weathering desert pavement results from the long term accumulation of fines and the weathering of surface clasts. This long term weathering includes; temperature change\ insolation, moisture and salt weathering. On UAE fluvial surfaces salt weathering appears to be only a minor process. This is in contrast to pavements in playa environments, such as Lake Eyre in Southern Australia, which are dominated by granular disintegration induced by salt weathering. This study shows that physical weathering has different rates on different rock types. Amongst pavements on similar rock type pavement properties can be used as surface age indicators.

ACKNOWLEDGMENTS

I would like to acknowledge Athari Al Shuwaihi for her help with the figures. I'm also grateful to Adrian Harvey for his feedback on this work.

REFERENCES

Al-Farraj, A., 1996. Late Pleistocene Geomorphology in Wadi Al-Bih northern UAE and Oman: with special emphasis on wadi terrace and alluvial fans. PhD. Thesis, University of Liverpool, 363pp.

Al-Farraj, A., 2002. Quaternary geomorphic evolution of wadi drainage basins in the northern UAE. *Bulletin of the Egyptian Geographical Society*, 75, 189-221.

Al-Farraj, A., 2008. Desert pavement development rate and processes on lake shorelines: case study of Lake Eyre (South), South Australia, *Geomorphology*, 100, 154-163.

Al-Farraj, A.M. and Harvey, A.M., 2000. Desert pavement characteristics on wadi terrace and alluvial fan surface: Wadi Al-Bih U.A.E. and Oman, *Geomorphology*, 35 279-297

Amit, R., Gerson, R., 1986. The Evolution of Holocene Reg (gravelly) soils in deserts - an example from the Dead Sea region. *Catena*, 13, 59-79.

Amit, R., Gerson, R., Yaalon, D.H., 1993. Stages and rate of gravel shattering process by salts in desert Reg soil. *Geoderma*, 57, 295-324.

Breed, C.S., McCauley, J.F., and Whitney, M.I., 1997. Wind erosion forms, in Thomas, D., ed., Arid zone Geomorphology: London, Belhaven Press, 437-464.

Cooke, R.U., 1970. Stone pavements in deserts. *Annals Association of American Geographers*, 60, 560-577.

Cooke, R. U., Warren, A., Goudie, A., 1993. Desert Geomorphology, London, UCL Press, 526 pp.

Cooke, R.U., Warren, A., 1973, Geomorphology in Deserts, London, 322pp.

Drake, N.A., Heydeman, M.T., and White, K.H., 1993. Distribution and formation of rock varnish in southern Tunisia. *Earth Surface Processes and Landforms*, 18, 31-41.

Goudie, A., 1997, Weathering processes, *in* Thomas, D., ed., Arid zone geomorphology: London, Belhaven, 11-24.

Guralnik,B., Matmon, A., Avni, Y., Fink, D., 2010, 10 Be exposure ages of ancient desert pavements reveal Quaternary evolution of theDead Sea drainage basin and rift margin tilting. *Earth and Planetary Science Letters*, 290, 132–141.

McCauley, J.F., Breed, C.S., El-Baz, F., Whitney, M.I., Grolier, M.J., Ward, A.W., 1979. Pitted and fluted rocks in the Western Desert of Egypt – Viking comparisons. *Journal of Geophysical research*, 84, 8222-8232.

McDonald, A.V., Pierson, F.B., Flerchinger, G.N., McFadden, L. D., 1996, Application of a soil-water balance model to evaluate the influence of Holocene climatic change on calcic, Mojave Desert, California, U.S.A., *Geoderma*, 74, 167-192.

McFadden, L.D., Wells, S.G., Jercinovich, M.J., 1987. Influences of eolian and pedogenic processes on the origin and evolution of desert pavements. *Geology*, 15, 504-508.

McFadden, L.D., Ritter, J.B., Wells, S.G., 1989. Use of multiparameter relative-age methods for estimation and correlation of alluvial fan surfaces on a desert piedmont, eastern Mojave Desert, California. *Quaternary research*, 32, 276-290.

McFadden, L.D., McDonald, E.V., Wells, S.G., Anderson, K., Quade, J., Forman, S.L., 1998. The Vesicular layer and carbonate collars in desert soils and pavement: formation, age and relation to climate change. *Geomorphology*, 24, 101-145.

McFadden, L.D., Eppes, M.C., Gillespie, A.R., Hallet, B., 2005. Physical weathering in Arid landscapes due to diurnal variation in the direction of solar heating. *Geological Society of America*, 117, no 1/2, 161-173.

Misak, R., 2000. Wind-Blown Features. In: El-Baz, F., and Al-Sarawi, M., (ed) Atlas of The State of Kuwait From Satellite Images. Boston University, 35-38.

Ollier, C.D., 1963. Insolation weathering: an example from central Australia. *American Journal of Science*, 261, 376-381.

Ollier, C.D., 1965. Dirt cracking- a type of insolation weathering. *Journal of Geology*, 73, 454-468.

Paton, T.R., Humphreys, G.S., Mitchell, P.B., 1995. Soils A new global view. UCL Press, 213p.

Springer, M.E., 1958. Desert pavement and vesicular layer of some soils of the Lahontan Basin, Nevada. *Soil Science Society of America Proceedings*, 22, 63-66.

Symons, P.M., Hemmings, C.F., 1968. A note on wind stable stone mantles in the southern Sahara. *Geographical Journal*, 134, 60-64.

Wainwright, J., Parsons, A.J., Abrahams, A.D., 1995. A Simulation study of the role of the erosion in the formation of the desert pavements. *Earth Surface Processes and Landforms*. 20,277-291.

Walles, S. G., McFadden, L.D., Poths, J., Olinger, C. T., 1995. Cosmogenic 3He surface-exposure dating of stone pavement: implication for landscape evolution in deserts. *Geology*, 23, 613-616.

Williams, S., Greeley, R., 1984. Desert pavement study at Amboy, California. *Planetary Geological Program*, 169-170.

Yaalon, D.H., 1970. Parallel stone cracking a weathering process on desert surfaces. *Bucharest, Geological Institute of Technical Economics Bulletin*, 18, 107-110.

INDEX

A

B

C

D

E

F

J

K

L

M

N

O

P

Q

R

S

T

U

V

W

X

Y

Z